V. Torra (Ed.)

Information Fusion in Data Mining

Dear MyCopy Customer,

This Springer book is a monochrome print version of the eBook to which your library gives you access via SpringerLink. It is available to you at a subsidized price since your library subscribes to at least one Springer eBook subject collection.

Please note that MyCopy books are only offered to library patrons with access to at least one Springer eBook subject collection. MyCopy books are strictly for individual use only.

You may cite this book by referencing the bibliographic data and/or the DOI (Digital Object Identifier) found in the front matter. This book is an exact but monochrome copy of the print version of the eBook on SpringerLink.

Springer-Verlag Berlin Heidelberg GmbH

Studies in Fuzziness and Soft Computing, Volume 123

http://www.springer.de/cgi-bin/search_book.pl?series=2941

Editor-in-chief
Prof. Janusz Kacprzyk
Systems Research Institute
Polish Academy of Sciences
ul. Newelska 6
01-447 Warsaw
Poland
E-mail: kacprzyk@ibspan.waw.pl

Further volumes of this series can be found on our homepage

Vol. 104. F. Rothlauf
Representations for Genetic and Evolutionary Algorithms", 2002
ISBN 3-7908-1496-2

Vol. 105. J. Segovia, P.S. Szczepaniak and M. Niedzwiedzinski (Eds.)
E-Commerce and Intelligent Methods, 2002

ISBN 3-7908-1499-7
Vol. 106. P. Matsakis and L.M. Sztandera (Eds.)
Applying Soft Computing in Defining Spatial Relations", 2002
ISBN 3-7908-1504-7

Vol. 107. V. Dimitrov and B. Hodge
Social Fuzziology, 2002
ISBN 3-7908-1506-3

Vol. 108. L.M. Sztandera and C. Pastore (Eds.)
Soft Computing in Textile Sciences, 2003
ISBN 3-7908-1512-8

Vol. 109. R.J. Duro, J. Santos and M. Graña (Eds.)
Biologically Inspired Robot Behavior Engineering, 2003
ISBN 3-7908-1513-6

Vol. 110. E. Fink l. 112. Y. Jin
Advanced Fuzzy Systems Design and Applications, 2003
ISBN 3-7908-1523-3

Vol. 111. P.S. Szcepaniak, J. Segovia, J. Kacprzyk and L.A. Zadeh (Eds.)
Intelligent Exploration of the Web, 2003
ISBN 3-7908-1529-2

Vol. 112. Y. Jin
Advanced Fuzzy Systems Design and Applications, 2003
ISBN 3-7908-1537-3

Vol. 113. A. Abraham, L.C. Jain and J. Kacprzyk (Eds.)
Recent Advances in Intelligent Paradigms and Applications", 2003
ISBN 3-7908-1538-1

Vol. 114. M. Fitting and E. Orowska (Eds.)
Beyond Two: Theory and Applications of Multiple Valued Logic, 2003
ISBN 3-7908-1541-1

Vol. 115. J.J. Buckley
Fuzzy Probabilities, 2003
ISBN 3-7908-1542-X

Vol. 116. C. Zhou, D. Maravall and D. Ruan (Eds.)
Autonomous Robotic Systems, 2003
ISBN 3-7908-1546-2

Vol 117. O. Castillo, P. Melin
Soft Computing and Fractal Theory for Intelligent Manufacturing, 2003
ISBN 3-7908-1547-0

Vol. 118. M. Wygralak
Cardinalities of Fuzzy Sets, 2003
ISBN 3-540-00337-1

Vol. 119. Karmeshu (Ed.)
Entropy Measures, Maximum Entropy Principle and Emerging Applications, 2003
ISBN 3-540-00242-1

Vol. 120. H.M. Cartwright, L.M. Sztandera (Eds.)
Soft Computing Approaches in Chemistry, 2003
ISBN 3-540-00245-6

Vol. 121. J. Lee (Ed.)
Software Engineering with Computational Intelligence, 2003
ISBN 3-540-00472-6

Vol. 122. M. Nachtegael, D. Van der Weken, D. Van de Ville and E.E. Kerre (Eds.)
Fuzzy Filters for Image Processing, 2003
ISBN 3-540-00465-3

Vicenç Torra (Ed.)

Information Fusion in Data Mining

Prof. Vicenç Torra
Institut d'Investigació
en Intel•ligència Artificial
CSIC - Spanish Scientific Research Council
Campus Universitat Autònoma de Barcelona
08193 Bellaterra, Catalonia
Spain
e-mail: vtorra@iiia.csic.es

ISSN 1434-9922

DOI 10.1007/978-3-540-36519-8

Library of Congress Cataloging-in-Publication-Data applied for

A catalog record for this book is available from the Library of Congress.

Bibliographic information published by Die Deutsche Bibliothek
Die Deutsche Bibliothek lists this publication in the Deutsche Nationalbibliographie; detailed bibliographic data is available in the internet at <http://dnb.ddb.de>.

http://www.springer.de

Originally published by Springer-Verlag Berlin Heidelberg New York in 2003
MyCopy version of the original edition 2003

Cover design: E. Kirchner, Springer-Verlag, Heidelberg
Printed on acid free paper 62/3020/M - 5 4 3 2 1 0
www.springer.com/mycopy

Preface

Information fusion is becoming a major need in data mining and knowledge discovery in databases. Due to this need, the interest on information fusion techniques is increasing in the data mining community. For example, typical applications of these techniques include the preprocessing step or data modeling (*e.g.* ensemble methods). Recent conferences (*e.g. Int. Conf. on Information Fusion* organized by the *Information Fusion society*) and special sessions or suggested topics in some *IEEE* organized conferences (*e.g. Annual Conf. of the Industrial Electronics Society, Workshop on Neural Networks for Signal Processing*) confirm this trend.

Nevertheless, the gap between both data mining and information fusion areas is large. While there are some information fusion methods currently in use in data mining (e.g. for remote sensing or image analysis and image fusion), some other tools are not much used in the field. Also, information fusion basic research and, specially, studies on formal methods for fusion, is usually more oriented to other kind of applications (e.g. decision making, tracking).

This book intends to be just a brick for the bridge to close this gap. To do so, we present some fusion techniques that are currently in use or could be used in data mining. On the other hand, we present some data mining application that use information fusion.

The book, after an introduction that outlines and classifies the main uses of information fusion techniques in data mining, is divided into four parts. The first part is devoted to aggregation operators, existing methods and their properties. This part contains three chapters.

In the first chapter, Torra reviews some of the existing aggregation operators for numerical data. He concentrates on the aggregation operators that belong to Choquet and Sugeno integral families (operators that are particular cases of these integrals).

Then, in the chapter by Narukawa and Murofushi, the authors study formal properties of the Choquet and Sugeno integrals. In particular, they show that under some conditions an aggregation operator can be expressed as a Choquet integral (or a Sugeno integral) for some particular fuzzy measure.

The next chapter, by Mitchell, describes the Probabilistic Weighted Ordered Weighted Average (PWOWA) operator. The author gives physical interpretation of the operator parameters. The paper demonstrates its applicability with an example on lossless image compression.

The second part deals with applications where information fusion techniques are applied for preprocessing. This part consists on three chapters.

First, Zhong studies data mining in the case of having multiple data sources. In particular, it studies how to extract the so-called "peculiarity rules" from multiple databases. The process is first illustrated with data concerning a supermarket sales database and latter on the method is applied to amino-acid data.

Then, in his paper, Tsumoto presents a method for knowledge discovery of temporal knowledge. The method is applied to medical databases. The method uses the so-called extended moving average method for removing noise and making temporal sequences smoother.

The last chapter of this part, by Torra and Domingo-Ferrer, reviews record-linkage methods and compares probabilistic and distance-based ones. These methods are applied for linking records that correspond to the same individual but belong to different data bases.

The third part is about using information fusion techniques for model building. Four chapters are included in this part.

In the first chapter, Grabisch reviews recent results on modeling data by means of Choquet integral and describes the application of genetic algorithms to this problem. The chapter also reviews some applications described in the literature where Choquet integral has been applied.

The next chapter by Imai, Asano and Sato also deals with the process of modeling using Choquet integrals. In particular, the authors present an algorithm for identifying fuzzy measures from examples. The algorithm is based on convex projections by Boyle and Dykstra.

Then, Hamuro, Katoh, Ip, Cheung and Yada introduce in their chapter a method for predicting future purchase behavior. The method extracts knowledge combining information fusion techniques (e.g. majority voting) with string pattern analysis. The approach is illustrated with several real-world examples related with customer purchase behavior.

Nakashima and Nakai close this part with a chapter that describes the use of fusion techniques to combine classification results. They consider multiple fuzzy rule-based systems for classification problems and given an input pattern combine the output of the systems to compute the pattern class. Their approach has been applied to several databases from the UCI repository.

Finally, the book finishes with a part devoted to the use of information fusion techniques for information extraction. This paper contains a chapter by Yager that introduces the use of linguistic summaries to fuse and present the information from a text database. The paper discusses how these summaries can be built from the database.

Sabadell, Catalonia
December, 2002

Vicenç Torra

Contents

Trends in Information fusion in Data Mining

Vicenç Torra

Institut d'Investigació en Intel·ligència Artificial - CSIC
Campus UAB s/n, E-08193 Bellaterra (Catalunya, Spain)

Abstract. This chapter reviews the main uses of information fusion techniques in the field of data mining. A classification of these uses is given into three rough classes: preprocessing, building models and information extraction.

1 Data Mining

Large amounts of data are nowadays available to companies, industries and researchers because gathering data is easy and usually inexpensive. However, most data is raw and to be useful relevant knowledge has to be extracted from it. Data Mining (DM) and Knowledge Discovery in Databases (KDD) are fields that study and provide methods for extracting this knowledge.

Data mining uses information fusion techniques for improving the quality of the extracted knowledge. Three main uses can be distinguished:

Information fusion in preprocessing: Fusion is used to increase the quality of raw data prior to the application of data mining methods.

Information fusion for building models: The model built from data uses some kind of information fusion technique (*e.g.* a particular aggregation operator to fuse partial results).

Information fusion is used to extract information: The knowledge extracted from data is the result of a particular information fusion technique. E.g., an aggregated value computed from the data.

In the next sections we discuss in more detail these three classes. Figure 1 outlines these classes.

2 Preprocessing data

Raw data, as collected by companies (*e.g.* with information about costumers and suppliers) or by sensors (*e.g.* with earth observed data) typically include some kind of erroneous data. Missing values and noisy data is the usual erroneous data found in files although in some cases inconsistent or distorted on purpose (e.g., protected by National Statistical Offices[5]) data is also found.

In this situation, fusion techniques can be applied to improve the quality of the data. This process is usually referred as *data cleaning* (or *data cleansing*).

Fusion methods are applied to those data referring to the same object / quality but either gathered by different information sources or by the same source but at different time instants. Alternatively, in some situations, data has to be fused (integrated) with previously established models. In this case, sometimes the process requires the revision of previously established models.

Depending on the available data (and of the data model, if any), fusion is done either at a low level or at a higher level. Tipically, lower levels require simple methods, usually for numerical (*e.g.* arithmetic mean, weighted mean and related aggregation operators) or linguistic/categorical data (majority or plurality rule – eventually weighted). Instead, when fusion is done at higher levels, more complex operators and formalisms are required. For example, evidence theory and logical approaches are used for model revision.

Naturally, methods are based on different assumptions about the data to be fused. The most basic need for sensory data is its registration: Recall that "*Sensor registration* refers to any of the means used to make the data from each sensor commensurate in both its spatial and temporal dimensions, i.e., that the data refer to the same location in the environment over the same period of time [13]". In fact, this need also appears in other application other than sensory data and becomes an important issue in any information fusion process.

Re-identification is a technique related, in a broad sense, with data registration. It consists on identifying data corresponding to the same object (*e.g.* individual, company or household). Record linkage, one of the particular methods for achieving re-identification, consists on linking records in separate files that relate to the same object. Re-identification is a required step when data mining tools have to be applied to distributed databases (multi database mining). In such situations, when non-homogeneous distributed databases are considered, databases are not fully consistent (either due to errors or because data has been protected to prevent disclosure risk).

> "For example, information on customers and suppliers is often distributed over several departments; data is stored in different platforms; data is not standardized (*e.g.* names and addresses have been written, and sometimes shortened, using different conventions). Under these circumstances, the result of a query can be incorrect (or inconsistent with previous results) as databases are not complete and they do not satisfy entity integrity." [18]

Re-identification and, more specifically, record linkage methods are applied to palliate the inconsistency on the results. These tools (*e.g.*, Winkler [21] implementation of probabilistic record linkage or Integrity commercial software [10]) are based on Statistical and Artificial Intelligence techniques. They use these technique for determining the matching between records (*e.g.*, probabilistic matching or fuzzy matching) and for extracting a unique identifier (or a set of variables acting as an identifier).

1. **Preprocessing data.**
   (a) Fusion to improve the quality of the data
   (b) Re-identification procedures
2. **Model building.**
   (a) Data models using aggregation operators
   (b) Aggregation operators to fuse data models
3. **Information extraction:**
   (a) Summaries and representatives of the data
   (b) Parameters of the aggregation operators

Fig. 1. Data fusion in data mining.

3 Model building

The main goal of data mining is to extract knowledge from data. This knowledge is usually represented by means of a particular data model that is extracted from the database. The set of alternative models considered in the literature is very large. Some of them are the following ones: regression analysis, rule based classifiers, neural networks. Several machine learning algorithms have been developed to learn these models from examples. Information fusion methods can be used in the process of model building.

In fact, information fusion techniques can be used in two ways.

On the one hand, fusion techniques can be used to define the model. This is, a particular aggregation operator is used for combining a set of inputs to obtain a given output. Let $\mathbb{C}$ ($\mathbb{C}$ from Consensus) be an aggregation operator (a function that takes N values and computes a kind of mean), then $\mathbb{C}$ takes data from N information sources (let $x_1, \ldots, x_N$ represent the sources) and computes an aggregated value. If we denote the data supplied by the information source x_i by $a_i = f(x_i)$, then, the aggregated value is:

$$\mathbb{C}_{param(\mathbb{C})}(a_1, \ldots, a_N)$$

where $\mathbb{C}_{param(\mathbb{C})}$ expresses that the operator $\mathbb{C}$ depends on some parameter and that the type of this parameter depends on $\mathbb{C}$.

Under these assumptions, given a set of M examples, where each example for $i \in \{1, \cdots, M\}$ consists on the values for N variables $(a_1^i a_2^i \ldots a_N^i)$ and the *correct* outcome b^i, the goal is to build a model of this data using an aggregation operator $\mathbb{C}$. Here, as usual, building a model consists on finding the operator $\mathbb{C}$ and their parameters $param(\mathbb{C})$ so that $\mathbb{C}_{param(\mathbb{C})}(a_1^i, \ldots, a_N^i)$ is as much similar as possible to b^i for all the examples.

In fact, the use of aggregation operators to build data models relies on the proposition proven in [17] that hierarchies of quasi-arithmetic means are universal approximators. In particular, let A denote the matrix $A = \{a_j^i\}$ and b be the vector such that $b' = (b^1 \ldots b^M)$. Then, it is said that when the original multi-dimensional input data $\mathbf{A}$ is extended to $(\mathbf{A}, -\mathbf{A}, -1, 0, 1)$, there exists a hierarchy of quasi-arithmetic means that approximates **b** to the required degree.

According to this, given some input data **A** and output **b**, it is, in principle, possible to build a data model of this data using aggregation operators. Naturally, different operators have different properties and different modeling capabilities being the hierarchies of quasi-arithmetic mean with inputs $(\mathbf{A}, -\mathbf{A}, -1, 0, 1)$ the more general model. At present, this is an active line of research. A short account of recent works is given in [19].

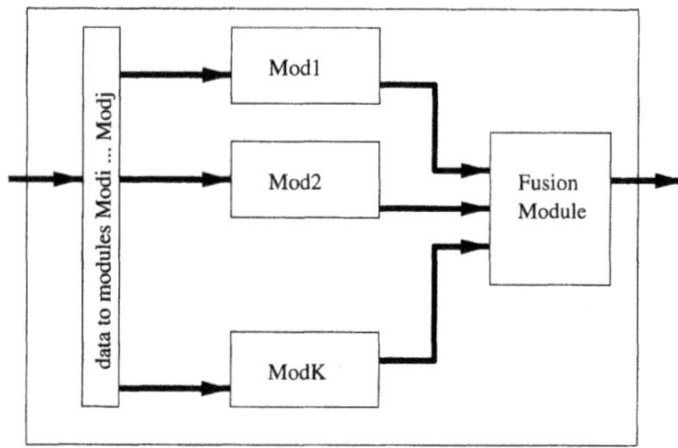

Fig. 2. Architecture to combine multiple models

On the other hand, information fusion can be used to combine several data models. This is the approach, for example, in [11], [14] and [20]. In general, the models built follow the architecture given in Figure 2: K alternative data models $Mod_1, \ldots, Mod_K$ (built from the data either using different methods or using the same method but applying it to different sets of examples or using different parameterizations) and an additional fusion module with a procedure to combine the output of the other modules. The operators considered in these combination modules include voting [14],[1] and the weighted mean [15]. Bagging [2] and Boosting [16] are well-known examples of machine learning algorithms for learning such complex data models.

4 Information extraction

A third use of information fusion in data mining is for extracting information. This is, for example, using information fusion techniques to build summaries or a kind of representatives from the original data. In fact, dimensionality reduction methods can also be seen from this perspective. The following situations can be considered in this case:

Reduce the number of records: Considering a data base with M records, this is to reduce the number of records to $M' < M$. Clustering, and more specifically methods for building cluster's *centroids* (either for numerical or categorical data[9,6]), is one of the methods for reducing the number of records.

Build summaries: A summary that includes relevant information is built from the original records. For example, [4] uses aggregation operators to built a per shot summary for video sequences (in a multimedia application). The process of building summaries can be seen as an extreme case of the reduction of the number of records.

Reduce the dimensionality of the records: Considering a database as a set of N dimensional records, this situation correspond to build a new set of N' dimensional records with $N' < N$ (with $N' = 1$ being the extreme case). Sammon's map (see e.g. [12], [8]), Principal Components Analysis or other well-known statistical techniques for dimensionality reduction (see *e.g.* [3]) are examples of methods for achieving this type of reduction.

The parameters of the aggregation operators offer an additional tool for information extraction. Due to the fact that a particular parameterization can be interpreted on the light of the parameters semantics (e.g., weights in weighted mean as a kind of relevance; weights in OWA operators in terms of orness or compensation; measures in Choquet integral in terms of interactions), when a data model is learned from data (*i.e.*, the model is built), the parameters can give valuable information about the data. See [7] for an application in this direction.

5 Conclusion

This chapter has given an overview of the main uses of information fusion techniques in the field of data mining. A classification of these uses was given. Figure 1 illustrates this classification.

References

1. Bauer, E., Kohavi, R., (1999), An Empiricial Comparison of Voting Classification Algorithms: Bagging, Boosting and Variants, Machine Learning 36 105-139.
2. Breiman, L., (1996), Bagging predictors, Machine Learning, 24 123-140.
3. Cox, T. F., Cox, M. A. A., (1994), Multidimensional scaling, Chapman and Hall.
4. Detyniecki, M., (2000), Mathematical Aggregation Operators and their Application to Video Querying, PhD dissertation, University of Paris VI, Paris, France.
5. Doyle, P., Lane, J. I., Theeuwes, J. J. M., Zayatz, L. M., (Eds.), (2001), Confidentiality, Disclosure, and Data Access: Theory and Practical Applications for Statistical Agencies, Elsevier.
6. Godo, L., Torra, V., (2000), On aggregation operators for ordinal qualitative information, IEEE T. on Fuzzy Systems, 8:143-154.

7. Grabisch, M., (2000), Fuzzy integral for classification and feature extraction, in M. Grabisch, T. Murofushi and M. Sugeno (Eds), Fuzzy Measures and Integrals, Physica-Verlag, 415-434.
8. Hastie, T., Tibshirani, R., Friedman, J., (2001), The Elements of Statistical Learning, Berlin: Springer.
9. Huang, Z., Ng., M. K., (1999), A fuzzy k-modes algorithm for clustering categorical data, IEEE Trans. on Fuzzy Systems, 7:4 446-452.
10. http://www.integrity.com
11. Ishibuchi, H., Morisawa, T., Nakashima, T., (1996), Voting Schemes for fuzzy-rule-based classification systems, Proc. of the Sixth IEEE Int. Conference on Fuzzy Systems, 614-620, Barcelona, Catalonia, Spain.
12. Kohonen, T., (1997), Self-Organizing maps, 2nd edition, Springer-Verlag.
13. Luo, R.C., Kay, M.G., (1992), Data fusion and sensor integration: State-of-the-art 1990s, in M. Al Abidi, R. C. Gonzalez, (Eds.), Data Fusion in Robotics and Machine Intelligence, Academic Press, 7-135.
14. Merz, C. J., (1999), Using Correspondence Analysis to Combine Classifiers, Machine Learning, 36 33-58.
15. Merz, C. J., Pazzani, M. J., (1999), Combining regression estimates, Machine learning, 36 9-32.
16. Schapire, R. E., (1990), The strength of weak learnability, Machine learning, 5:2 197-227.
17. Torra, V., (1999), On Some Relationships between Hierarchies of Quasiarithmetic Means and Neural Networks, Int. J. of Intel. Syst. 14:11 1089–1098.
18. Torra, V., (2000), Towards the re-identification of individuals in data files with non-common variables, Proc. of the European Conf. on Artificial Intelligence (ECAI 2000), 326-330, Berlin, Germany.
19. Torra, V., (2003), Information Fusion in Data Mining: Outline, Chapter in this book.
20. Webb, G.I., (2000), MultiBoosting: A Technique for Combining Boosting and Wagging, Machine Learning, 40 159-196.
21. Winkler, W. E., (1995), Advanced methods for record linkage, American Statistical Association, Proc. of the Section on Survey Research Methods, 467-472.

Part 1:

Aggregation Operators: Methods and Properties

On some aggregation operators for numerical information

Vicenç Torra

Institut d'Investigació en Intel·ligència Artificial - CSIC
Campus UAB s/n, 08193 Bellaterra (Catalonia, Spain)
e-mail: vtorra@iiia.csic.es, http://www.iiia.csic.es/~vtorra

Abstract. In this chapter we review some of the aggregation operators that are appropriate for fusing numerical information. We focus on the ones that belong to Choquet and Sugeno integral families. This is, the ones that these integrals generalize. In particular, the review includes, among others, the following operators: arithmetic mean, weighted mean, OWA and WOWA operators, weighted minimum and weighted maximum, Sugeno and Choquet integral.

1 Introduction

Knowledge integration is an essential process in any intelligent system. These systems have to take into account that information sources (*e.g.*, sensors or experts) supply partial and, sometimes, erroneous information. Due to this fact, systems are forced to consider (integrate or fuse) data provided by several sources.

Information fusion is the field that studies all the aspects related to the combination of information. It studies the aggregation functions (aggregation operators), the properties that these functions satisfy and how to build a function from a set of imperative properties. It could be said that the goals of this field are to systematize (and formalize) the synthesis process and to characterize the existing methods. Both aspects are needed if it is intended to apply combination functions to new problems.

Information fusion techniques are applied in several fields like mathematics, economy, biology. Different methods have been developed according to the type of information to be fused[1] and according to the properties these methods have to satisfy. At present, there are functions operating on several formalisms, e.g., numerical and categorical data (either in nominal or ordinal scales), probability and possibility distributions, fuzzy sets, quantitative and qualitative preference relations, clustering results (in several formalisms: dendrograms, partitions), ...

In this work we review the aggregation functions for numerical values. We focus on two main families of operators: the ones that are generalized

[1] In fact, for each type of object there is at least a fusion method that can be applied: the plurality function that returns the most voted element

by Choquet and Sugeno integrals. These two families encompass most of well-known aggregation operators. The structure of the paper is as follows. Section 2 reviews these operators, and then Section 3 summarizes some recent research on model building and, more specifically, on learning parameters for some of the operators previously described. The chapter finishes with some conclusions.

2 The fuzzy t-conorm integral family

This section is devoted to the fuzzy t-conorm integral family. Section 2.1 reviews, among others, the weighted mean and the OWA operator. Then, Section 2.2, after introducing fuzzy measures, describes the Choquet integral and relates it with the previously described operators. Section 2.3 is about the Sugeno integral family. Definitions in this section include, *e.g.* the weighted minimum, the weighted maximum and the Sugeno integral itself. Finally, Section 2.4 describes the fuzzy t-conorm integral.

2.1 The choquet integral family

The arithmetic mean and the weighted mean are probably the most widely used aggregation functions. These functions that operate with numerical values together with the quasi-arithmetic mean (a general form), were studied and characterized in [1,3]. Recently, another function, the OWA operator, was defined [54,55] to model the aggregation of data in intelligent systems.

Both functions, the weighted mean and the OWA operator, are a lineal combination of the values according to a vector of weights. Their difference is on the semantics (the meaning) of the weights in each function. On the one hand, the weighted mean allows the system to compute an aggregate value from the ones corresponding to several information sources taking into account the reliability of each source. In fact, each source has an attached weight that measures the reliability or expertise of the data it supplies in the final decision. On the other hand, the OWA operator permits of weighting the values in relation to their ordering. In this way, for example, when modelling a decision maker we can reduce (or ignore the influence of) extreme values. We consider below the definition of both aggregation operators. For the sake of completeness, the definition of the arithmetic mean is also given.

Definition 1. A vector $v = (v_1 ... v_N)$ is a *weighting vector* of dimension N if and only if $v_i \in [0,1]$ and $\sum_i v_i = 1$.

Definition 2. A mapping $AM\colon \mathbb{R}^N \to \mathbb{R}$ is an *arithmetic mean* of dimension N if $AM(a_1, ..., a_N) = (1/N) \sum_i a_i$.

Definition 3. Let $\mathbf{p}$ be a weighting vector of dimension N, then a mapping $WM\colon \mathbb{R}^N \to \mathbb{R}$ is a *weighted mean* of dimension N if $WM_{\mathbf{p}}(a_1, ..., a_N) = \sum_i p_i a_i$.

Definition 4. [54] Let $\mathbf{w}$ be a weighting vector of dimension N, then a mapping OWA: $\mathbb{R}^N \to \mathbb{R}$ is an *Ordered Weighting Averaging (OWA) operator* of dimension N if

$$OWA_{\mathbf{w}}(a_1, ..., a_N) = \sum_{i=1}^{N} w_i a_{\sigma(i)}$$

where $\{\sigma(1), ..., \sigma(N)\}$ is a permutation of $\{1, ..., N\}$ such that $a_{\sigma(i-1)} \geq a_{\sigma(i)}$ for all $i = \{2, ..., N\}$ (i.e. $a_{\sigma(i)}$ is the i-th largest element in the collection $a_1, ..., a_N$).

In the statistics literature, the OWA operator is known (see [33]) by the *L-estimators* (a linear combination of order statistics). Therefore, the OWA generalizes all order statistics.

It is easy to proof that when the weighting vector of both WM and OWA is such that $v = (1/N, ..., 1/N)$ then, both operators reduce to the arithmetic mean (AM). This is $WM_v = OWA_v = (1/n) \sum_i a_i$.

According to what was said, the meaning of the two weighting vectors $\mathbf{p}$ and $\mathbf{w}$ above is different. We give below some examples of applications that can illustrate this difference (we use below $\mathbf{p}$ as the weights in the weighted mean and $\mathbf{w}$ as the weights in the OWA operator):

- In multicriteria decision making (or also when evaluating solutions in constraint satisfaction problems [44]), the weights $\mathbf{p}$ correspond to the importance of the experts or the criteria to combine, while the weights $\mathbf{w}$ correspond to compensation among values.
- In robot sensing when data is recorded in the same time instant, weights p_i correspond to the reliability of the i-th sensor, while the weights $\mathbf{w}$ correspond to our evaluation of the importance of small values (or large values). For example, we can give a greater weight to small values than to large ones if our system can collide with an object (*i.e.*, $w_1 > w_2 > w_3 > \ldots$).
- Also in robot sensing, when samples have been obtained at different instants of time, the weights $\mathbf{p}$ allow the system to give more importance to recent data while weights $\mathbf{w}$ can model the fact that previous data is rellevant when it has sensed a near object.

These examples have shown that weights have a different semantics in both combination functions and then they can be used for different purposes. However, although both points of view are meaningful in a single problem, both combination functions present the drawback of considering only one of them. To solve this drawback, in [42] was defined a combination function that allows a system to consider at the same time both the relevance of the information sources (the $\mathbf{p}$-weights) and the relevance of the values (the $\mathbf{w}$-weights). This function is the WOWA (Weighted OWA) operator.

Definition 5. [41] Let $\mathbf{p}$ and $\mathbf{w}$ be two weighting vectors of dimension N, then a mapping $WOWA$: $\mathbb{R}^N \to \mathbb{R}$ is a *Weighted Ordered Weighted Averaging (WOWA) operator* of dimension N if

$$WOWA_{\mathbf{p},\mathbf{w}}(a_1, ..., a_N) = \sum_{i=1}^{N} \omega_i a_{\sigma(i)}$$

where σ is defined as in the case of the OWA (i.e., $a_{\sigma(i)}$ is the i-th largest element in the collection $a_1, ..., a_N$), and the weight ω_i is defined as:

$$\omega_i = w^*(\sum_{j \le i} p_{\sigma(j)}) - w^*(\sum_{j<i} p_{\sigma(j)})$$

with w^* being a monotonic increasing function that interpolates the points $(i/N, \sum_{j \le i} w_j)$ together with the point $(0,0)$. The function w^* is required to be a straight line when the points can be interpolated in this way.

[45], [4] and [50] deal about the method for building the interpolated function w^*. [51] contains an implementation of WM, OWA and WOWA operators. This operator can be directly defined using the function w^*. In such case, it is equivalent to the OWA operator with importances [56] because w^* is a non-decreasing function and, therefore, can be interpreted as a non-decreasing fuzzy quantifier. We give below the WOWA for a weighting vector $\mathbf{w}$ and a fuzzy quantifier Q:

Definition 6. A function $Q : [0,1] \to [0,1]$ is a *regular monotonically non-decreasing fuzzy quantifier* (non-decreasing fuzzy quantifiers for short) if it satisfies: (i) $Q(0) = 0$; (ii) $Q(1) = 1$; (iii) $x > y$ implies $Q(x) \ge Q(y)$.

Definition 7. Let Q be a non-decreasing fuzzy quantifier, and let $\mathbf{w}$ be a weighting vector of dimension N, then a mapping WOWA: $\mathbb{R}^N \to \mathbb{R}$ is a *Weighted Ordered Weighted Averaging (WOWA) operator* of dimension N if

$$WOWA_{\mathbf{p},Q}(a_1, ..., a_N) = \sum_{i} \omega_i a_{\sigma(i)}$$

where σ is defined as in the case of the OWA, and the weight ω_i is defined as:

$$\omega_i = Q(\sum_{j \le i} p_{\sigma(i)}) - Q(\sum_{j<i} p_{\sigma(i)})$$

This operator is defined keeping the semantics of the weights $\mathbf{p}$ and $\mathbf{w}$ as it is in the weighted mean and in the OWA. Besides of that it can be proven that this operator is a generalization of the other two.

Proposition 1. *[41] The WOWA operator reduces to a weighted mean when $w_i = 1/N$ and reduces to an OWA operator when $p_i = 1/N$.*

Several generalizations exist for these operators. Besides of the Choquet integral that will be reviewed in Section 2.2 it is important to mention the quasi-arithmetic mean (and quasi-weighted mean – also called quasi-linear mean [13], quasi-OWA, ...). A quasi-arithmetic mean applies a mapping f to each value a_i and then reverses the mapping once the aggregated value is obtained. This is:

Definition 8. Let f be an invertible function from $\mathbb{R}^+$ to $\mathbb{R}^+$, then a mapping QAM: $(\mathbb{R}^+)^N \to \mathbb{R}^+$ is a *quasi-arithmetic mean* of dimension N if

$$QAM(a_1, ..., a_N) = f^{-1}((1/N) \sum_i f(a_i))$$

Selecting appropriate functions f, the quasi-arithmetic mean reduces to some well-known aggregation functions. For example, $f(x) = Kx+K'$ reduces to an arithmetic mean, $f(x) = K \ln x + K'$ reduces to the geometric mean and $f(x) = Kx^{-1} + K'$ reduces to the harmonic mean. This family and the quasi-weighted mean were studied and characterized in [1] and [2]. [13] (p.134) defined the quasi-OWA and [42] the quasi-WOWA.

2.2 Aggregation using fuzzy measures: the Choquet integral

Now, we consider a different aggregation operator: the Choquet integral. The main difference between this operator and the previous ones from the conceptual point of view is the consideration of fuzzy measures. In the weighted mean, we combine the values to aggregate with a weighting vector, and these weights are defined for each element. I.e., for each a_i in $\{a_1, \ldots, a_N\}$ we have its corresponding p_i or $w_{\sigma(i)}$. In fact, strictly speaking, if $X = \{x_1, \ldots, x_N\}$ are the information sources, p_i is associated to x_i. So, p_i is the importance or reliability of source (expert / sensor / criteria) x_i. However, up to now, we have not considered the importance of sets of elements. At present, if we consider the importance (the measure of the importance) of a set of elements (i.e., a set of sensors or experts), we can only define it as the addition of the importance of the individual elements. This is, $p(\{x_1, x_2, x_3\}) = p_1 + p_2 + p_3$. However, in fact other measures can be considered.

A classical example in the literature is the following one [14]. We are to evaluate a set of students in relation to three subjects: {Mathematics, Physics, Literature}. We want to give more importance to science-related subjects than to literature, but on the other hand we want to give some advantage to students that are good both in literature and in any of the science related subjects. A simple way to model this is with a fuzzy measure $\mu : \wp(X) \to [0, 1]$. This is, we define the importance for any subset of subjects as a value in $[0, 1]$. For example, we can define a measure according to the previous guidelines in the following way [14]:

1. Boundary conditions:

$\mu(\emptyset) = 0$, $\mu(\{M, P, L\}) = 1$
The importance of the empty set is zero. The set X has maximum importance.
2. Relative importance of scientific versus literary subjects:
$\mu(\{M\}) = \mu(\{P\}) = 0.45$, $\mu(\{L\}) = 0.3$
The importance of mathematics and physics is larger than the importance of literature.
3. *Redudancy* between mathematics and physics:
$\mu(\{M, P\}) = 0.5 < \mu(\{M\}) + \mu(\{P\})$
Mathematics and physics are similar subjects. The importance of the set containing both should not be larger than the addition.
4. Support between literature and scientific subjects:
$\mu(\{M, L\}) = \mu(\{P, L\}) = 0.9 > \mu(\{P\}) + \mu(\{L\})$
Mathematics and literature are complementary subjects.

As in the above example, fuzzy measures are appropriate when there is redundancy / complementariness among sources or when information sources are not independent. In such situations, we expect that the aggregation operators fuse the information taking into account such redundancy / complementariness. Some interpretations of fuzzy measures are given in [30] and a review of several indices for measuring the interactions between sources is given in [17].

In general, a fuzzy measure is a set function that satisfies boundary conditions and monotonicity. The measure is defined over the set of information sources X. Therefore, it gives some information (background knowledge following the terminology in Artificial Intelligence) about the sources.

Definition 9. (see [53]) A fuzzy measure μ on a set X is a set function $\mu : \wp(X) \rightarrow [0, 1]$ satisfying the following axioms:

(i) $\mu(\emptyset) = 0$, $\mu(X) = 1$ (boundary conditions)
(ii) $A \subseteq B$ implies $\mu(A) \leq \mu(B)$ (monotonicity)

We want to underline that a fuzzy measure replaces the axiom of additive measures ($\mu(A \cup B) = \mu(A) + \mu(B)$ when $A \cap B = \emptyset$) by a more general one: monotonicity. Thus, probability measures are also fuzzy measures. In relation to this, we see that weighting vectors in a weighted mean can be understood as being equivalent to additive fuzzy measures. This is clear when we define $\mu(C) = \sum_{x_i \in C} \mu(\{x_i\})$ where $\mu(\{x_i\})$ is the weights corresponding to the i-th information source: p_i using the notation given above.

In fact, besides of being able to classify fuzzy measures according to their additivity and non-additivity, there exists a large set of fuzzy measure families. Some of them are the following ones: Sugeno λ-measures [38], $\perp$-decomposable ones (see *e.g.* [20]), hierarchically decomposable ones [47], k-order additive ones [15]. All of them are particular cases of the general ones defined above and they are adequate to express knowledge when some restrictions about the information sources are fullfilled. We consider below some examples of these families.

In fact, one of the motivations for defining such families is that the definition of a fuzzy measure requires 2^N parameters (recall that a fuzzy measure is a set function). Therefore, for real applications with a large number of information sources, the number of values to be supplied is very large. Families of fuzzy measures with restricted complexity are appropriate in such situations because the number of parameters are much less. For example, N values in Sugeno λ-measures; N values and $(N-1)$ t-conorms[2] (or $(N-1)$ parameters if a parametric family of t-conorms is selected) in hierarchically decomposable ones; $N+(N-1)^2$ in 2-order additive fuzzy measures.

Definition 10. A fuzzy measure μ on a set X is a $\perp$-decomposable fuzzy measure if there exists a t-conorm $\perp$ such that for all $A, B \subseteq X$ with $A \cap B \neq \emptyset$ it holds:

$$\mu(A \cup B) = \mu(A) \perp \mu(B)$$

Definition 11. A fuzzy measure μ on a set X is a *Q-p-decomposable fuzzy measure* if there exists a non-decreasing fuzzy quantifier Q and a weighting vector $\mathbf{p} : X \to [0,1]$ such that for all $C \subseteq X$ it holds:

$$\mu(C) = Q(\sum_{x_i \in C} p(x_i))$$

These measures are also called distorted probabilities in *e.g.* [22]. In fact, a fuzzy measure μ is a distortion of fuzzy measure ν when $\nu(A) \leq \nu(B)$ implies $\mu(A) \leq \mu(B)$. See [23] for basic references.

Decomposable or Q-p-decomposable fuzzy measures are used when the measure of a set is a function of the importance of the elements that define this set. These definitions show that the measure can be defined with only $|X|$ values and a t-conorm or a fuzzy quantifier, instead of requiring a $2^{|X|}$ values as needed for an arbitrary fuzzy measure.

The introduction of fuzzy measures raises the interest on aggregation operators that use this kind of information in the fusion process. At present, several operators with such capability have been defined. Examples include Choquet and Sugeno integral. The Choquet integral is defined below and the Sugeno integral is defined in Section 2.3

Definition 12. [8] Let μ be a fuzzy measure on X, then the *Choquet integral* of a function $f : X \to \mathbb{R}^+$ with respect to the fuzzy measure μ is defined by:

$$(C) \int f d\mu = \sum_{i=1}^{N} [f(x_{s(i)}) - f(x_{s(i-1)})] \mu(A_{s(i)}) \qquad (1)$$

[2] t-conorms are monotonic, commutative and associative ($[0,1] \times [0,1] \to [0,1]$ operators with neutral element 0. They are used in fuzzy logic for modeling disjunction (fuzzy set union). See *e.g.* [27] for details

where $f(x_{s(i)})$ indicates that the indices have been permuted so that $0 \leq f(x_{s(1)}) \leq \cdots \leq f(x_{s(N)}) \leq 1$, $A_{s(i)} = \{x_{s(i)}, \cdots, x_{s(N)}\}$ and $f(x_{\sigma(0)}) = 0$.

A Choquet integral can be expressed in alternative ways according to the next proposition.

Proposition 2. *(see* e.g. *[20]) Let μ be a fuzzy measure on X, then the Choquet integral of a function $f : X \to \mathbb{R}^+$ with respect to μ can be expressed as:*

$$(C)\int f d\mu = \sum_{i=1}^{N} f(x_{\sigma(i)})[\mu(A_{\sigma(i)}) - \mu(A_{\sigma(i-1)})] \tag{2}$$

or as:

$$(C)\int f d\mu = \sum_{i=1}^{N} f(x_{s(i)})[\mu(A_{s(i)}) - \mu(A_{s(i+1)})] \tag{3}$$

where $\{\sigma(1), \cdots, \sigma(N)\}$ is a permutation of $\{1, \cdots, N\}$ such that $f(x_{\sigma(1)}) \geq f(x_{\sigma(2)}) \geq \cdots \geq f(x_{\sigma(N)})$, $A_{\sigma(k)} = \{x_{\sigma(j)} | j \leq k\}$ (or, equivalently, $A_{\sigma(k)} = \{x_{\sigma(1)}, \cdots, x_{\sigma(k)}\}$ when $k \geq 1$ and $A_{\sigma(0)} = \emptyset$), and where s and $A_{s(i)}$ are as in Definition 12.

This latter proposition shows clearly that a Choquet integral is a kind of weighted mean of the values $f(a_{\sigma(i)})$ with the weights $p_{\sigma(i)} = \mu(A_{\sigma(k)}) - \mu(A_{\sigma(k-1)})$. Note that $p_{\sigma(i)} \geq 0$ and that $\sum p_{\sigma(i)} = 1$. The former definition shows that in a Choquet integral each segment $f(x_{s(i)}) - f(x_{s(i-1)})$ is considered (weighted) according to all the elements x_j such that $f(x_{s(j)}) \geq f(x_{s(i)})$. This is, the importance of each segment $(f(x_{s(i)}) - f(x_{s(i-1)}))$ corresponds to the measure of all the elements whose evaluation embeds that segment (i.e., x_j such that $f(x_{s(j)}) \geq f(x_{s(i)})$).

For details on interpretations of the Choquet integral and comparison with the Lebesgue integral see [33]. Some properties of the integral and a characterization is given in [34].

Although the Choquet integral is defined with respect to fuzzy measures, and although the aggregation operators previously considered are defined with respect to weighting vectors, it can be proved that both methods are related. In fact, the WM, the OWA and the WOWA operators are Choquet integrals for particular fuzzy measures. The following theorems establish these relationships.

Theorem 1. *[12], [32] For every weighting vector* $\mathbf{p}$*, we have* $WM_{\mathbf{p}} = CI_{\mu}$ *with* μ *defined by:*

$$\mu_{\mathbf{p}}(B) = \sum_{x_i \in B} p_i \qquad \text{for all } B \subseteq X$$

This theorem shows that the Choquet integral is a proper generalization of weighted mean for non additive measures. This is, when a measure is additive, Choquet integral reduces to a weighted mean.

Theorem 2. *[12], [32] For every weighting vector* $\mathbf{w}$*, we have* $OWA_{\mathbf{w}} = CI_\mu$ *with* μ *defined by:*

$$\mu_{\mathbf{w}}(B) = \sum_{i=1}^{|B|} w_i \qquad \text{for all } B \subseteq X$$

Theorem 3. *[43] For every weighting vector* $\mathbf{p}$ *and every regular monotonically non-decreasing fuzzy quantifier* Q*, we have* $WOWA_{\mathbf{p},Q} = CI_\mu$ *with* μ *defined by:*

$$\mu_{\mathbf{p},Q}(B) = Q(\sum_{x_i \in B} p_i) \qquad \text{for all } B \subseteq X$$

Due to the fact that for WOWA operatorswith two weighting vectors, we can define $WOWA_{\mathbf{w},Q}$, this implies that there exists a fuzzy measure μ such that $WOWA_{\mathbf{p},w} = CI_\mu$.

These theorems show that the weighted mean, OWA and WOWA operators are particular cases of Choquet integrals. There are some results that establish the reversal conditions. This is, when a Choquet integral can be reduced to the former operators. For example, additive measures lead to WM and commutative measures (*i.e.*, $\mu(A) = \mu(B)$ when $|A| = |B|$) lead to OWA operators.

In this section we have given a Choquet integral definition that restricts numerical data to be in $\mathbb{R}^+$. This is, $f(x_i) \geq 0$ for all x_i in X. Two extensions of Choquet integrals are of interest for dealing with negative values: the symmetric integral (or Šipoš integral) and the asymmetric integral. They are described and compared in [18].

2.3 The Sugeno integral family

There exists an alternative group of aggregation operators that are not directly related with the WM, the OWA and the WOWA operators nor the Choquet integral. They are the so-called weighted min and the weighted max. Their definition is given below. Note that although they consider also a weighting vector, this vector is not equivalent to the one given previously in Definition 1. Here it is required that at least one of the weights is one, and the addition of all the weights can be greater than one. We will call such weighting vector a ***possibilistic*** weighting vector (because it is equivalent to have a possibility distribution [27] on the set $X = \{x_1, ..., x_N\}$). The former vector (in Definition 1) will be referred by ***probabilistic*** weighting vector.

Definition 13. A vector $v = (v_1 ... v_N)$ is a *possibilistic weighting vector* of dimension N if and only if $v_i \in [0, 1]$ and $\max_i v_i = 1$.

Definition 14. [9] Let $\mathbf{u}$ be a weighting vector of dimension N, then a mapping $WMin$: $[0, 1]^N \to [0, 1]$ is a *weighted minimum* of dimension N if $WMin_{\mathbf{u}}(a_1, ..., a_N) = \min_i \max(neg(u_i), a_i)$.

Definition 15. [9] Let $\mathbf{u}$ be a weighting vector of dimension N, then a mapping $WMax$: $[0,1]^N \to [0,1]$ is a *weighted maximum* of dimension N if $WMax_{\mathbf{u}}(a_1, ..., a_N) = \max_i \min(u_i, a_i)$.

These aggregation operators cannot be directly related with the ones described in previous sections. In fact, only in some particular cases the values coincide. For example: (1) the OWA coincides with the weighted min when both return the minimum value (i.e., $w = (0, ..., 0, 1)$ and $u = (1, ..., 1)$) and (2) the OWA coincide with the weighted max when both return the maximum value (i.e., $w = (1, 0, ..., 0)$ and $u = (1, ..., 1)$).

Weighted maximum can be seen as having a definition analogous to the weighted mean replacing product by minimum and addition by maximum. In fact, this can be understood on the light of fuzzy logic, then a t-conorm (product) is replaced by another one (minimum) and a t-conorm (bounded sum) is replaced by another t-conorm (maximum). Naturally, the same approach can also be applied to the OWA operators, obtaining the so-called OWMax. This approach is just adding the ordering process to the weighted maximum. Applying the same process to weighted minimum we get the so-called OWMin. Both definitions are given below.

Definition 16. [10] Let $\mathbf{u}$ be a weighting vector of dimension N, then a mapping $OWMin$: $[0,1]^N \to [0,1]$ is an *ordered weighted minimum* of dimension N if:

$$OWMin_{\mathbf{u}}(a_1, ..., a_N) = \min_i \max(neg(u_i), a_{\sigma(i)})$$

where σ is a permutation as in Definition 4

Definition 17. [10] Let $\mathbf{u}$ be a weighting vector of dimension N, then a mapping $OWMax$: $[0,1]^N \to [0,1]$ is an *ordered weighted maximum* of dimension N if:

$$OWMax_{\mathbf{u}}(a_1, ..., a_N) = \max_i \min(u_i, a_{\sigma(i)})$$

where σ is a permutation as in Definition 4

Now we consider the Sugeno integral. In the same way that the Choquet integral generalizes the OWA and the WM, the Sugeno integral generalizes the WMin and the WMax. Its definition is given below:

Definition 18. [38] Let μ be a fuzzy measure on X, then the *Sugeno integral* (SG) of a function $f : X \to [0,1]$ with respect to μ is defined by:

$$(S)\int f d\mu = \max_{i=1,N} \min(f(x_{s(i)}), \mu(A_{s(i)})) \tag{4}$$

where $f(x_{s(i)})$ indicates that the indices have been permuted so that $0 \leq f(x_{s(1)}) \leq ... \leq f(x_{s(N)}) \leq 1$, $A_{s(i)} = \{x_{s(i)}, ..., x_{s(N)}\}$ and $f(x_{s(0)}) = 0$.

The next proposition states that the Sugeno integral generalizes the $WMin$ and the $WMax$ and defines the fuzzy measure so that the equality with SG is possible.

Proposition 3. *[9] Sugeno integral generalizes both weighted minimum and weighted maximum:*

1. *A weighted maximum with a possibilistic weighting vector* $\mathbf{u}$ *is equivalent to a Sugeno integral with the fuzzy measure:*

$$\mu_{\mathbf{u}}^{wmax}(A) = \max_{a_i \in A} u_i$$

2. *A weighted minimum with a possibilistic weighting vector* $\mathbf{u}$ *is equivalent to a Sugeno integral with the fuzzy measure:*

$$\mu_{\mathbf{u}}^{wmin}(A) = 1 - \max_{a_i \notin A} u_i$$

Therefore, $WMax_{\mathbf{u}} = SG_{\mu_{\mathbf{u}}^{wmax}}$ *and* $WMin_{\mathbf{u}} = SG_{\mu_{\mathbf{u}}^{wmin}}$.

Interpretations of Sugeno integrals with respect to fuzzy measures are given in [53] and [58]. A chapter in this book [34] studies some of the properties of Sugeno integrals and gives a characterization.

2.4 Fuzzy t-conorm integrals

Although only in some particular situations the Choquet integral and the Sugeno integral have the same outcome for the same input, there exist a more general integral, the fuzzy t-conorm integral, that generalizes both. The fuzzy t-conorm integral is defined over a tuple called a t-conorm system for integration, and an operation $-_{\Delta}$ based on one of the elements of this tuple. We give now their corresponding definitions and the one of the fuzzy t-conorm integral.

The t-conorms presented here and the product like operation $\otimes$ define, when some conditions are fulfilled, a t-conorm system.

Definition 19. [31] $\mathcal{F} = (\Delta, \perp, \underline{\perp}, \otimes)$ is a *t-conorm system for integration* if and only if:

1. $\Delta, \perp, \underline{\perp}$, are continuous t-conorms, which are the maximum or Archimedean.
2. $\otimes : ([0,1], \Delta) \times ([0,1], \perp) \rightarrow ([0,1], \underline{\perp})$ is a product-like operation fulfilling:
   (a) $\otimes$ is continuous on $(0,1]^2$
   (b) $a \otimes x = 0$ if and only if $a = 0$ or $x = 0$
   (c) when $x \perp y < 1$, then $a \otimes (x \perp y) = (a \otimes x) \underline{\perp} (a \otimes y)$ for all $a \in [0,1]$
   (d) when $a \Delta b < 1$, then $(a \Delta b) \otimes x = (a \otimes x) \underline{\perp} (b \otimes x)$, for all $x \in [0,1]$.

In this definition $([0,1], \varDelta), ([0,1], \perp), ([0,1], \underline{\perp})$ correspond, respectively, to the spaces of values of integrand, measure and integral; and letters k, g and h express generators of the t-conorms $\varDelta, \perp, \underline{\perp}$ when they exists (e.g., $\varDelta, \perp, \underline{\perp}$ are continous Archimedean)

According to the definition there are four types of t-conorm systems [31]:

Type (i): $\varDelta, \perp, \underline{\perp}$ are Archimedean t-conorms.
Type (ii): $\varDelta, \perp, \underline{\perp}$ are the maximum.
Type (iii): $\underline{\perp}$ is an Archimedean t-conorm, and at least one of the others is the maximum.
Type (iv): $\underline{\perp}$ is the maximum, and at least one of the others is Archimedean.

However, although these four types can be considered the analysis of their properties show that only t-conorm systems (i) and (ii) are meaningful (see Propositions 2.4 and 2.5 in [31] for details). Therefore, here we will focus on types (i) and (ii).

Definition 20. [31] For a given t-conorm $\varDelta$, we define an operation $-_{\varDelta}$ on $[0,1]^2$ by: $a -_{\varDelta} b := inf\{c | b \varDelta c \geq a\}$.

In our case, the following two cases for $-_{\varDelta}$ are of interest:

1. When $\varDelta$ is an Archimedean t-conorm with a generator k, then

$$a -_{\varDelta} b = k^{-1}[max(0, k(a) - k(b))]$$

2. When $\varDelta$ is the maximum, then

$$a -_{\varDelta} b = \begin{cases} a \text{ if } a \geq b \\ 0 \text{ if } a < b \end{cases}$$

Definition 21. [31] Let $(X, \mathcal{X}, \mu)$ be a fuzzy measure space and $\mathcal{F} = (\varDelta, \perp, \underline{\perp}, \otimes)$ be a t-conorm system for integration. For a measurable function $f : X \to [0,1]$, the fuzzy t-conorm integral (or fuzzy t-integral) of f based on $(\varDelta, \perp, \underline{\perp}, \otimes)$ with respect to μ is defined as:

$$(\mathcal{F}) \int f \otimes d\mu = lim_{n \to \infty} (\underline{\perp}) \int f_n \otimes d\mu$$

where $\{f_n\}$ is a non-decreasing sequence of simple functions which pointwise converges to f.

In the case of a discrete domain, an alternative expression, more convenient from the computational point of view, can be found. This equivalent expression is given below. The definition assumes that $\mathcal{X} = \wp(X)$.

Proposition 4. *[31] Let μ be a fuzzy measure on X, let $\mathcal{F} = (\Delta, \perp, \underline{\perp}, \otimes)$ be a t-conorm system for integration. Then, the fuzzy t-conorm integral (or fuzzy t-integral) of a function $f : X \to [0,1]$ based on $(\Delta, \perp, \underline{\perp}, \otimes)$ with respect to μ as defined above is equivalent to:*

$$(\mathcal{F}) \int f \otimes d\mu = \underline{\perp}_{i=1}^{N} (a_i -_{\Delta} a_{i-1}) \otimes \mu(A_{s(i)})$$

where $a_i = f(x_{s(i)})$, $a_0 = f(x_{s(0)}) = 0$ and $A_{s(i)} = \{x_{s(i)}, \cdots, x_{s(N)}\}$.

It has been said that fuzzy t-conorm integrals generalize both the Choquet integral and the Sugeno integral. Note that:

1) the fuzzy t-conorm integral with the t-conorm system $(\hat{+}, \hat{+}, \hat{+}, \cdot)$ where $\hat{+}$ stands for the t-conorm $\hat{+}(x, y) = x + y - xy$ and $\cdot$ stands for the ordinary product, corresponds to the Choquet integral

2) the fuzzy t-conorm integral with the t-conorm system (Max, Max, Max, Min) corresponds to the Sugeno integral.

Although the fuzzy t-conorm integral generalizes both the Choquet integral and the Sugeno one, not all possible t-conorm systems generate fuzzy t-conorm integrals that are adequate for aggregation. Among all the t-conorm systems of types (i) and (ii) only those with $\underline{\perp} = \Delta$ satisfy properties that are considered a requirement for aggregation. This is so because, as pointed out by [20], the spaces of the integral $([0,1], \Delta)$ and integrand $([0,1], \underline{\perp})$ must be the same if the integral is seen as a mean value of the integrands. In this case, when fuzzy t-conorm integrals are generated by t-conorm systems of the form $(\Delta, \perp, \Delta, \otimes)$ with Δ and $\perp$ being archimedean t-conorm they are called restricted fuzzy t-conorm integrals, instead when they are generated by t-conorm systems of the form $(Max, Max, Max, \otimes)$ they are called quasi-Sugeno integrals.

Restricted fuzzy t-conorm integrals and quasi-Sugeno integrals satisfy the following properties (see Section 6.8.2 in [20] and [26] for details and proofs):

1) They satisfy unanimity. This is, when $f(x) = a$ for all $x \in X$ the results of the integrals are a.

2) They are monotonic in relation to the measurable function f and the fuzzy measure μ. This is, if $f(x) \leq f'(x)$ for all $x \in X$ and if $\mu(B) \leq \mu'(B)$ for all $B \in X$ then the following inequalities hold:

$$(\mathcal{F}) \int f \otimes d\mu \leq (\mathcal{F}) \int f' \otimes d\mu$$

$$(\mathcal{F}) \int f \otimes d\mu \leq (\mathcal{F}) \int f \otimes d\mu'$$

3) The value of the integral is within the infimum and the supremum of the values of $f(x)$. This is, the following inequalities hold (this is a continuous counterpart of the usual condition for aggregation operators that the aggregated value is within the Minimum and the Maximum of the values to aggregate):

$$inf f \leq (\mathcal{F}) \int f \otimes d\mu \leq sup f$$

Note that the Choquet integral and the Sugeno integral, as particular cases of restricted ones, satisfy all these properties.

3 Model building

A major research problem in the field of aggregation operators is the development of algorithms and methodologies to determine the suitable model for a certain problem. This corresponds to both selection (i) selection of the most suitable aggregation technique and (ii) determination of corresponding parameters. Although the characterization of the methods can help in this process, development of methodologies and software is neeeded to help in such determination. Specially, software is needed for parameter determination.

At present there exist several methods for parameter determination. One of the most well-known methods is Saaty's Analytical Hierarchy Process (AHP) [37]. An interview-based method for the weighted mean. The method assumes that a user can be interviewed and from his/her answers WM weights are extracted. A similar approach was developed by O'Hagan [36] and later on further developed by Carbonell et al. in [7]. This approach, applied to OWA operators, also requires an expert to supply some information. In this case, the information is the so-called *orness* of the OWA. Orness is a measure of the degree of similarity between the fused value and the maximum of the input values (i.e., maximum orness – orness equal to 1 – makes OWA equal to $\max a_i$ and minimum orness – orness equal to 0 – makes OWA equal to $\min a_i$). O'Hagan method selects the weights that for a given orness maximize their dispersion (dispersion maximization is to ensure that parameter sensitiveness is minimal). Therefore, a value for orness that can be understood as a compensation measure [44] completely determines OWA weights. A related approach but for determining fuzzy measures (distorted probabilities) is described in [22].

A completely alternative approach is when such external *expert* is not available and, instead, there are a set of examples. Aggregation operators are then built as approximations of the available data. Therefore, the goal is to determine the model from the examples. The usual case is that examples are pairs of (input, output) pairs following the structure in Table 1. This is, each row $((a_1^i, a_2^i, \ldots, a_N^i), b^i)$ is an example. This approach has been considered for learning the parameters of several aggregation operators.

Tanaka and Murofushi in [40] considered the learning of fuzzy measures for Choquet integral. The same subject, under different assumptions, has been considered in [29], [20] and [24]. Chapters by Grabisch [19] and Imai [25] in this book also deal with the learning of fuzzy measures for Choquet integral. Filev and Yager studied in [11] the determination of OWA weights from examples. An alternative approach was considered in [48] for the OWA and WM. The method had a better performance and was applied to larger sets with a larger number of variables. [49] considered the learning of the weights for the WOWA operator (equivalent to learning distorted probabilities) using a combination of the techniques in [11] and [48]. This latter approach was compared with another based on genetic algorithms in [35]. Recently, learning for the quasi-weighted mean has been considered. In [52] parameters for quasi-weighted means are determined. Still, to reduce the complexity of the problem, the invertible function f is assumed to be of restricted form.

Table 1. Data examples.

$$\begin{array}{cccc|c} a_1^1 & a_2^1 & \dots & a_N^1 & b^1 \\ a_1^2 & a_2^2 & \dots & a_N^2 & b^2 \\ \vdots & \vdots & & \vdots & \vdots \\ a_1^M & a_2^M & \dots & a_N^M & b^M \end{array}$$

4 Conclusions

In this Chapter we have briefly reviewed some of the numerical aggregation operators and their relationship. We have focused on the aggregation operators that belong to the Choquet and Sugeno integral family. We have also looked over the literature for parameter determination.

Additional information about aggregation operators can be found in the recently edited volumes about aggregation operators [5] (this book includes an extensive survey [6]) and fuzzy measures and integrals [21]. Marichal in [28] presents a good and detailed overview of operators, properties and indices. Other reference books about aggregation and related topics include [13], [20] and [39]. More specific topics are dealt in *e.g.* [53] (fuzzy measure theory) and [57] (OWA operators).

Acknowledgment

Partial support of the European Community under the contract "CASC" IST-2000-25069 and of the Spanish Ministry of Science and Technology under the project "STREAMOBILE" (TIC2001-0633-C03-02) is acknowledged. This chapter is a revised and expanded version of [46].

References

1. Aczél, J., (1984), On weighted synthesis of judgements, Aequationes Math., 27 288-307.
2. Aczél, J., (1987), A short course on functional equations, D. Reidel Publishing Company.
3. Aczél, J., Alsina, C., (1987), Characterization of some classes of quasilinear functions with applications to triangular norms and to synthesizing judgements, Methods of Operations Research, 48.
4. Beliakov, G., (2001), Shape preserving splines in constructing WOWA operators, Fuzzy Sets and Systems, 113 389-396.
5. Calvo, T., Mayor, G., Mesiar, R., (Eds.), (2002), Aggregation operators: New Trends and Applications, Physica-Verlag.
6. Calvo, T., Kolesárová, A., Komorníková, M., Mesiar, R., (2002), Aggregation Operators: Properties, Classes and Construction Methods, in T. Calvo, G. Mayor, R. Mesiar (Eds.), Aggregation operators: New Trends and Applications, Physica-Verlag, 3-123.
7. Carbonell, M., Mas, M., Mayor, G., (1997), On a class of Monotonic Extended OWA Operators, Proc. of the Sixth IEEE International Conference on Fuzzy Systems (IEEE-FUZZ'97), 1695-1699, Barcelona, Catalonia, Spain.
8. Choquet, G., (1953), Theory of Capacities, Ann. Inst. Fourier 5 131-296.
9. Dubois, D., Prade, H., (1986), Weighted Minimum and Maximum Operations in Fuzzy Set Theory, Information Sci. 39 205-210.
10. Dubois, D., Prade, H., Testemale, C., (1988), Weighted fuzzy pattern-matching, Fuzzy Sets and Systems, 28 313-331.
11. Filev. D. P., Yager, R. R., (1998), On the issue of obtaining OWA operator weights, Fuzzy Sets and Systems, 94 157-169.
12. Fodor, J., Marichal, J. L., Roubens, M., (1995), Characterization of the Ordered Weighted Averaging Operators, IEEE Trans. on Fuzzy Systems, 3:2 236-240.
13. Fodor, J., Roubens, M., (1994), Fuzzy preference modelling and multicriteria decision support, Kluwer Academic Publishers.
14. Grabisch, M., (1995), Fuzzy integral in multicriteria decision making, Fuzzy Sets and Systems 69 279-298.
15. Grabisch, M., (1996), k-order additive fuzzy measures, Proc. 6th Int. Conf. on Information Processing and Management of Uncertainty in Knowledge-Based Systems (IPMU), 1345-1350, Granada, Spain.
16. Grabisch, M., (1998), Fuzzy Integral as a Flexible and Interpretable Tool of Aggregation, in B. Bouchon-Meunier, (Ed.), Aggregation and Fusion of Imperfect Information, Physica-Verlag, 51-72.
17. Grabisch, M., (2000), The Interaction and Möbius Representation of Fuzzy Measures on Finite Spaces, k-Additive Measures: A Survey, in M. Grabisch, T. Murofushi, M. Sugeno, (Eds.), Fuzzy Measures and Integrals: Theory and Applications, Physica-Verlag, 70-93.
18. Grabisch, M., (2002), Aggregation Based on Integrals: Recent Results and Trends, in T. Calvo, G. Mayor, R. Mesiar (Eds.), Aggregation operators: New Trends and Applications, Physica-Verlag, 107-123.
19. Grabisch, M., (2003), Modelling data by the Choquet integral, Chapter in this book.

20. Grabisch, M., Nguyen, H. T., Walker, E.A., (1995), Fundamentals of Uncertainty Calculi with Applications to Fuzzy Inference, Kluwer Academic Publishers, Dordreht, The Netherlands.
21. Grabisch, M., Murofushi, T., Sugeno, M., (Eds.), (2000), Fuzzy Measures and Integrals: Theory and Applications, Physica-Verlag.
22. Honda, A., Nakano, T., Okazaki, Y., (2002), Subjective evaluation based on distorted probability, Proc. of the SCIS & ISIS Conference (CD-ROM), Tsukuba, Japan.
23. Honda, A., Nakano, T., Okazaki, Y., (2002), Distortion of Fuzzy Measures, Proc. of the SCIS & ISIS Conference (CD-ROM), Tsukuba, Japan.
24. Imai, H., Miyamori, M., Miyakosi, M., Sato, Y., (2000), An algorithm Based on Alternative Projections for a Fuzzy Measures Identification Problem, Proc. of the Iizuka Conference, Iizuka, Japan (CD-Rom).
25. Imai, H., Asano, D., Sato, Y., (2003), An Algorithm Based on Alternative Projections for a Fuzzy Measure Identification Problem, Chapter in this book.
26. Kandel, A., Byatt, W. J., (1978), Fuzzy sets, fuzzy algebra, and fuzzy statistics, Proc. of the IEEE 66 1619-1639.
27. Klir, G. J., Yuan, B., (1995), Fuzzy Sets and Fuzzy Logic: Theory and Applications, Prentice-Hall, U.K.
28. Marichal, J.-L., (1999), Aggregation operators for multicriteria decision aid, PhD Dissertation, Faculte des Sciences, Universite de Liege, Liège, Belgium.
29. Marichal, J.-L., Roubens, M., (2000), Determination of weights of interacting criteria from a reference set, European Journal of Operational Research 124:3 641-650.
30. Murofushi, T., Sugeno, M., (1989), An interpretation of fuzzy measure and Choquet integral as an integral with respect to a fuzzy measure, Fuzzy Sets and Systems, 29 201-227
31. Murofushi, T., Sugeno, M., (1991), Fuzzy t-conorm integral with respect to fuzzy measures: generalization of Sugeno integral and Choquet integral, Fuzzy Sets and Systems, 42:1 57-71.
32. Murofushi, T., Sugeno, M, (1993), Some quantities represented by the Choquet integral, Fuzzy Sets and Systems, 56 229-235.
33. Murofushi, T., Sugeno, M., (2000), Fuzzy Measures and Fuzzy Integrals, in M. Grabisch, T. Murofushi, M. Sugeno (Eds.), Fuzzy Measures and Integrals: Theory and Applications, Physica-Verlag, 3-41.
34. Narukawa, Y., Murofushi, T., (2003), Choquet integral and Sugeno integral as aggregation functions, Chapter of this book.
35. Nettleton, D., Torra, V., (2001), A comparision of active set method and genetic algorithm approaches for learning weighting vectors in some aggregation operators, Int. J. of Intel. Syst., 16 1069-1083.
36. O'Hagan, M., (1988), Aggregating template or rule antecedents in real-time expert systems with fuzzy set logic, Proceedings of the 22nd Annual IEEE Asilomar Conference on Signals, Systems and Computers, 681-689, Pacific Grove, CA, USA.
37. Saaty, T. L., (1980), The Analytic Hierarchy Process, McGraw-Hill.
38. Sugeno, M., (1974), Theory of fuzzy integrals and its applications, Ph. D. dissertation, Tokyo Institute of Technology.
39. Sugeno, M., Murofushi, T., (1993), Fuzzy Measure, Tokyo, Nikkan Kogyo Shinbunsha (in Japanese).

40. Tanaka, A., Murofushi, T., (1989), A learning model using fuzzy measure and the Choquet integral, Proc. of the 5th Fuzzy System Symposium, 213-217, Kobe, Japan (in Japanese).
41. Torra, V., (1996), Weighted OWA operators for synthesis of information, Proc. of Fifth IEEE Int. Conference on Fuzzy Systems (IEEE-FUZZ'96) (ISBN 0-7803-3645-3), 966-971, New Orleans, USA.
42. Torra, V., (1997), The Weighted OWA operator, Int. J. of Intel. Syst., 12 153-166.
43. Torra, V., (1998), On some relationships between the WOWA operator and the Choquet integral, Proc. of the Seventh Int. Conf. on Information Processing and Management of Uncertainty in Knowledge-Based Systems (IPMU) (ISBN 2-84254-013-1), 818-824, Paris, France.
44. Torra, V., (1998), On considering constraints of different importance in fuzzy constraint satisfaction problems, Int. J. of Uncertain Fuzziness and Knowledge-Bases 6:5 489-501.
45. Torra, V., (2000), The WOWA operator and the interpolation function W*: Chen and Otto's interpolation method revisited, Fuzzy Sets and Systems, 113:3 389-396.
46. Torra, V., (1998), On the integration of numerical information: from the arithmetic mean to fuzzy integrals, Report de Recerca, DEI-RR-98-010, Tarragona, Universitat Rovira i Virgili.
47. Torra, V., (1999), On hierarchically S-decomposable fuzzy measures, Int. J. of Intel. Syst., 14:9 923-934.
48. Torra, T., (1999), On the learning of weights in some aggregation operators: The weighted mean and the OWA operators. Mathware and Soft Computing, 6 249-265.
49. Torra, V., (2000), Learning Weights for Weighted OWA Operators, Proc. of the IEEE Int. Conference on Industrial Electronics, Control and Instrumentation (IECON 2000), CD-ROM, Nagoya, Japan.
50. Torra, V., (2001), Authors's reply (to an article by G. Beliakov), Fuzzy Sets and Systems, 121 551.
51. Torra, V., (2002), http://www.iiia.csic.es/~vtorra/wowa/wowa.html
52. Torra, V., (2002), Learning weights for the quasi-weighted means, IEEE Trans. on Fuzzy Systems, 10:5 653-666.
53. Wang, Z., Klir, G., (1992), Fuzzy measure theory, Plenum Press.
54. Yager, R.R., (1988), On ordered weighted averaging aggregation operators in multi-criteria decision making, IEEE Trans. on SMC, 18 183-190.
55. Yager, R.R., (1993), Families of OWA operators, Fuzzy Sets and Systems, 59 125-148.
56. Yager, R. R., (1996), Quantifier Guided Aggregation Using OWA operators, Int. J. of Int. Syst., 11 49-73.
57. Yager, R. R., Kacprzyk, J., (1997), The Ordered Weighting Averaging Operators, Kluwer Academic Publishers.
58. Yoneda, M., Fukami, S., Grabisch, M., (1994), Human factor and fuzzy science, in K. Asai (Ed.), Fuzzy science, Kaibundo, 93-122 (in Japanese).

Choquet integral and Sugeno integral as aggregation functions

Yasuo Narukawa[1] and Toshiaki Murofushi[2]

[1] Toho Gakuen, 3-1-10, Naka, Kunitachi, Tokyo, 186-0004, Japan. narukawa@d4.dion.ne.jp

[2] Comp. Intell. & Syst. Sci., Tokyo Inst. Tech. 4259 Nagatsuta, Midoriku, Yokohama, 226-8502, Japan. murofusi@fz.dis.titech.ac.jp

Abstract. The basic properties of Choquet integral and Sugeno integral as aggregation tools are shown. The conditions for aggregation functions to be the Choquet integral or the Sugeno integral is presented. Selecting appropriate aggregation tool is fundamental for having a good modeling and thus studying their properties is a fundamental issue. It is also shown that the Choquet integral and Sugeno type integral are essential, from a purely mathematical view, in the generalized fuzzy integral.

1 Introduction

An aggregation function (an aggregation operator) is a functional from $[0,1]^n$ to $[0,1]$, where $[0,1]^n$ is a set of all functions $x : \{1, 2, \cdots, n\} \to [0,1]$. An aggregation function is used for building data models or for the information extraction [22]. The Choquet integral [3] and the Sugeno integral [20] , which are the generalizations of the max, min, mean and median operation, are commonly used as aggregation functions [6,7]. Selecting appropriate aggregation tool is fundamental for having a good modeling and thus studying their properties is a fundamental issue.

In this chapter, we present basic properties of Choquet integral and Sugeno integral as aggregation tools. We show the conditions for aggregation functions to be the Choquet integral or the Sugeno integral. We also show that the Choquet integral and Sugeno type integral are essential, from a mathematical view, in the generalized fuzzy integral, that are investigated in the various literature [9,13,17,21,23].

In Section 2, we present the basic properties of Choquet and Sugeno integral and we show these are the generalization of the max, min, mean and median operation.

In Section 3, we state about the generalized fuzzy integral based on [15]. We define a $t-$ conorm system for integration, $t-$ conorm integral and the generalized $t-$ conorm integral, that is a generalization of some other investigations. We show that , using additive generators, the essential integral, in the mathematical viewpoint, is only Choquet integral or Sugeno type integral.

In Section 4, we present conditions for aggregation functions to be the Choquet or Sugeno integral. Choquet integral and Sugeno integral cannot be used for a system without any conditions, even if they are quite general ones. We show the necessary and sufficient conditions for aggregation functions to be Choquet or Sugeno integral. It is important for using the Choquet integral and Sugeno integral in data mining to keep these conditions in mind.

2 Choquet integral and Sugeno integral

In this section, we introduce the Choquet integral and the Sugeno integral and present their basic properties.

2.1 Comonotonicity

Throughout this Chapter we assume that X is a finite set, that is, $X := \{1, 2, \cdots, n\}$. $[0,1]^n$ be the space of all nonnegative real function x from X to the unit interval $[0,1]$.

The $i-$th order statistic $x^{(i)}$ [26] is a functional on $[0,1]^n$ which is defined by arranging the components of $x = (x_1, \cdots, x_n) \in [0,1]^n$ in the increasing order

$$x^{(1)} \leq \cdots \leq x^{(i)} \leq \cdots \leq x^{(n)}.$$

Next, we present the definition of comonotonicity, which is the essential property for Choquet integral and Sugeno integral.

Definition 1 [4] Let $x, y \in [0,1]^n$.

We say that x and y are comonotonic if

$$x_i < x_j \Rightarrow y_i \leq y_j$$

for $i, j \in X$.

Let $x \in [0,1]^n$. We call the sets $\{i | x_i > r\}, \{i | x_i \geq r\}$ the strict and weak, respectively, upper level set of X for the level $r \in [0,1]$. The class M_x of all upper level sets of x is called the upper set system of x.

A chain of sets in 2^X is a set system $M \subset 2^X$ which is completely ordered with respect to set inclusion, i. e.

$$A, B \in M \text{ implies } A \subset B \text{ or } B \subset A.$$

Let P be a probability measure on $(X, 2^X)$ and $E_P(x)$ be the expectation of x, that is,

$$E_P(x) := \sum_{i \in X} x_i P(i).$$

Proposition 2 [1,5] *For $x, y \in [0,1]^n$ the following conditions are equivalent:*

1. *x, y are comonotonic.*
2. *$(x_i - x_j)(y_i - y_j) \geq 0$ for $i, j \in X$.*

3. *Let Mx, My be the upper set system of x, y, respectively. Then $Mx \cup My$ is a chain.*
4. *There exists a function $z : X \to [0,1]$ and increasing functions u, v on $[0,1]$ such that*
$$x = u(z)\ ,\ y = v(z)$$
.
5. *$E_P(x) \cdot E_P(y) \le E_P(xy)$, where $E_P(xy) := \sum_{i \in X} x_i y_i P(i)$.*

Next, we present the basic definitions of a fuzzy measure.

Definition 3 [20] A fuzzy measure μ on $(X, 2^X)$ is a real valued set function, $\mu : 2^X \longrightarrow [0,1]$ with the following properties. A triplet $(X, 2^X, \mu)$ is said to be a fuzzy measure space.

1. $\mu(\emptyset) = 0\ ,\ \mu(X) = 1$
2. $\mu(A) \le \mu(B)$ whenever $A \subset B$, $A, B \in 2^X$.

2.2 Choquet integral

In this subsection, we define the Choquet integral and present its basic properties. Choquet integral can be regarded as the expectation of $x \in [0,1]^n$ measured by the interval scale [10]. We show, as the example, which the Choquet integrals with respect to some fuzzy measure with a parameter can output the min. the max and the mean of the input $x \in [0,1]^n$ by changing the value of parameter.

Definition 4 [3,14] Let μ be a fuzzy measure on $(X, 2^X)$. The Choquet integral of $x \in [0,1]^n$ with respect to μ is defined by

$$C_\mu(x) := \int_0^\infty \mu_x(r)dr,$$

where $\mu_x(r) = \mu(\{x_i \ge r\})$.

Using the i–th order statistics, the Choquet integral is written as

$$C_\mu(x) = \sum_{i=1}^{n} (x^{(i)} - x^{(i-1)})\mu(\{(i) \cdots (n)\}),$$

where we define $x^{(0)} := 0$.

Theorem 5 [5] *Let μ be a fuzzy measure on $(X, 2^X)$, then the following expression hold:*

1. (positive homogeneity)
$$C_\mu(ax) = aC_\mu(x)$$
for $x \in X$ and a positive real number $a \in [0,1]$.

2. (monotonicity)

$$x \leq y \Rightarrow C_\mu(x) \leq C_\mu(y)$$

for $x, y \in R_+^n$, *where* $x \leq y$ *means* $x_i \leq y_i$ *for all* $i \in X$.

3. (comonotonic additivity)

$$C_\mu(x+y) = C_\mu(x) + C_\mu(y)$$

for comonotonic x *and* $y \in [0,1]^n$.

Especially we have

$$C_\mu(ax+b) = aC_\mu(x) + b$$

for $x \in [0,1]^n$ and constants $a, b \in [0,1]$. This means that the Choquet integral can be understood as the expectation of x measured in an interval scale.

Let $\mathcal{S}_n$ be the set of all permutation of X. For each $\pi \in \mathcal{S}_n$, we define

$$\mathcal{D}_\pi := \{x \in [0,1]^n | x_{\pi(1)} \leq \cdots x_{\pi(n)}\}.$$

It is obvious that each $\mathcal{D}_\pi$ is a polyhedron (i.e. an intersection of closed halfspaces) and $\bigcup_{\pi \in \mathcal{S}_n} \mathcal{D}_\pi = R^n$. It is obvious that x and $x' \in \mathcal{D}_\pi$, are comonotonic. Moreover, by Theorem 5, the restrictions of Choquet integral C_μ on $\mathcal{D}_\pi$ are monotone, positive homogeneous and additive.

The next example shows that the Choquet integrals with respect to $\lambda-$fuzzy measure[20] can output the min, the max and the mean of the input $x \in [0,1]^n$ by changing the value of parameter λ.

Example 1 Let P be a probability measure on $(X, 2^X)$. Define a function $\varphi_\lambda : R \to R$ by

$$\varphi_\lambda(r) := \log_{1+\lambda}(1+\lambda r)$$

and a fuzzy measure μ_λ by $\mu_\lambda := \varphi_\lambda^{-1} \circ P$. Then we have

$$\lim_{\lambda \to -1} C_{\mu_\lambda}(x) = \max_{i \in x} x_i$$

$$C_{\mu_0}(x) = E_P(x)$$

$$\lim_{\lambda \to \infty} C_{\mu_\lambda}(x) = \min_{i \in x} x_i$$

2.3 Sugeno integral

In this subsection, we define the Sugeno integral and show its basic properties. Sugeno integral is constructed using only the ordinal structure of the space. Hence the Sugeno integral can deal with the systems that can be measured by only ordinal scale [10]. The Sugeno integral can be regarded as a generalization of median operation. In this Chapter, $\wedge$ means the minimum operation, $\vee$ means the maximum operation.

Definition 6 [20] Let μ be a fuzzy measure on $(X, 2^X)$. The Sugeno integral of $x \in [0,1]^n$ with respect to μ is defined by

$$S_\mu(x) := \sup_{r \in [0,1]} [r \wedge \mu_x(r)]$$

where $\mu_x(r) := \mu(\{x_i \geq r\})$.

Using the i−th order statistics, the Sugeno integral is written as

$$S_\mu(x) = \bigvee_{i=1}^{n} (x^{(i)} \wedge \mu(\{(i) \cdots (n)\})).$$

Theorem 7 [2,20] *Let μ be a fuzzy measure on $(X, 2^X)$, then the following expression hold:*

1. (min homogeneity)
$$C_\mu(a \wedge x) = a \wedge C_\mu(x)$$
for $x \in X$ and a positive real number $a \in [0,1]$.
2. (monotonicity)
$$x \leq y \Rightarrow S_\mu(x) \leq S_\mu(y)$$
for $x, y \in [0,1]^n$.
3. (comonotonic maxitivity)
$$S_\mu(x \vee y) = S_\mu(x) \vee S_\mu(y)$$
for comonotonic x and $y \in [0,1]^n$.

It is well known that the Sugeno integral is a generalization of median operator

Theorem 8 [12] *Let μ be a fuzzy measure on $(X, 2^X)$ and $x \in [0,1]^n$. We have*

$$S_\mu(x) = \text{median}\{x_1, \cdots, x_n, \mu(\{(2), \cdots, (n)\}), \cdots, \mu(\{(i) \cdots (n)\}), \cdots, \mu(\{(n)\})\}.$$

3 Generalized fuzzy integral

In this section, we use $t-$conorms and $t-$norms. They are binary operators that generalize addition and multiplication, and also max and min. We show that, from a purely mathematical view, the essential integral operation are only Choquet integral and Sugeno type integral.

A triangular conorm (t-conorm) $\perp$ is a binary operation on $[0,1]$ fulfilling the conditions:

(T1) $x \perp 0 = x$

(T2) $x \perp y \leq u \perp v$ whenever $x \leq u$ and $y \leq v$.

(T3) $x \perp y = y \perp x$

(T4) $(x \perp y) \perp z = x \perp (y \perp z)$.

A t-conorm is said to be strict if and only if it is continuous on $[0,1]$ and strictly increasing in each of its places. A t-conorm $\perp$ is said to be Archimedean if and only if $x \perp x > x$ for all $x \in (0,1)$.

Example 2

1. The maximum operator $x \vee y$ is a non Archimedean t-conorm.
2. The bounded sum $x \hat{+} y := 1 \wedge (x+y)$ is an Archimedean t-conorm.
3. Sugeno operator $x +_\lambda y := 1 \wedge (x + y + \lambda xy)$ $(-1 < \lambda < \infty)$ is an Archimedean t-conorm.

Proposition 9 [11] *If a t-conorm $\perp$ is continuous and Archimedean, then there exists a continuous and strictly increasing function $g : [0,1] \to [0,\infty]$ such that $x \perp y = g^{(-1)}(g(x) + g(y))$, where $g^{(-1)}$ is the pseudo inverse of g which is defined by*

$$g^{(-1)}(u) := \begin{cases} g^{(-1)}(u) & \text{if } u \leq g(1) \\ 1 & \text{if } u > g(1). \end{cases}$$

The function g is called an additive generator of a t-conorm $\perp$.

Example 3 φ_λ in Example 1 is the additive generator of Sugeno operator $+_\lambda$.

Let m be a fuzzy measure on $(X, 2^X)$. We say that m is $\perp-$decomposable if $m(A \cup B) = m(A) \perp m(B)$ whenever $A \cap B \neq \emptyset$ for $A, B \in 2^X$.

The continuous t-conorms we consider in this chapter are restricted to Archimedean ones or the maximum operator. The first reason is the any non Archimedean continuous t-conorm is represented by Archimedean continuous t-conorms and $\vee$. This fact is mentioned by Weber [24]. The second reason is the decomposition theorem by Sugeno and Murofushi [21]: if $\perp$ is a non Archimedean t-conorm which is not $\vee$, the $\perp-$ decomposable fuzzy measure space $(X, 2^X, m)$ with the decomposability can be decomposed into decomposable fuzzy measure space $\{(X_i, \mathcal{X}_i, m_i)\}$ where $\{X_i\}$ is a partition of X, $\mathcal{X}_i := \{A \cap X_i | A \in \mathcal{X}\}$, m_i is $\perp_i$ decomposable measure and $\perp_i$ can be regarded as an Archimedean t-conorm or $\vee$.

Definition 10 A t-conorm system for integration (or t-system for short) is a quadruplet $(F, M, I, \boxdot)$ consisting of $F := ([0,1], \perp_1)$, $M = ([0,1], \perp_2)$, $I = ([0,1], \perp_3)$, where $\perp_i$ $i = 1,2,3$ are continuous t-conorms which are $\vee$ or Archimedean, and a non decreasing operator $\boxdot : F \times M \to I$ satisfying

(M1) $\boxdot$ is left continuous on $(0,1]$.

(M2) $a \boxdot x = 0$ if and only if $a = 0$ or $x = 0$.

(M3) if $x \perp_2 y < 1$ then $a \boxdot (x \perp_2 y) = (a \boxdot x) \perp_3 (a \boxdot y)$.

(M4) if $a \perp_1 b < 1$ then $(a \perp_1 b) \boxdot x = (a \boxdot x) \perp_3 (b \boxdot x)$.

The letters g_i express the generators of $\perp_i : i = 1,2,3$ if they exist.

The t-system is expressed by $(\perp_1, \perp_2, \perp_3, \boxdot) : i = 1,2,3$ instead of $(F, M, I, \boxdot)$. For example, a t-system is expressed by $(+_\lambda, \vee, \hat{+}, \cdot)$ in the case where $\perp_1 = +_\lambda, \perp_2 = \vee$, $\perp_3 = \hat{+}, \boxdot = \cdot$(the ordinary multiplication), and by $(\vee, \vee, \vee, \wedge)$ in the case where $\perp_1 = \perp_2 = \perp_3 = \vee$ and $\boxdot = \wedge$.

Example 4 A uninorm [25] is a binary operation on the unit interval which is commutative, associative, non-decreasing in each component, and which has a neutral element. Let $([0,1], S, U)$ be a conditionally distributive semiring [8]; that is, S is a continuous $t-$conorm and U is a left continuous uninorm satisfying the conditional distributivity of U over S: for all $x, y, z \in [0,1]$ with $S(x,y) < 1$ we have $U(x, S(y,z)) = S(U(x,y), U(x,z))$. Suppose that $U(0,x) = 0$ for $x \in [0,1]$. $(\perp_1, \perp_2, \perp_3, \boxdot)$ is a $t-$ system in the case where $\perp_1 = \perp_2 = \perp_3 = S$ and $\boxdot = U$.

Since we consider the Archimedean t-conorms or the maximum operator $\vee$, there are four types of $t-$system.

1. $\perp_i : i = 1,2,3$ are Archimedean.
2. $\perp_1 = \perp_2 = \perp_3 = \vee$
3. $\perp_3$ is Archimedean and at least one of the other is $\vee$.
4. $\perp_3 = \vee$ and at least one of the other is Archimedean.

Proposition 11 *Let $(\perp_i, \boxdot) : i = 1,2,3$ be a t-system.*

1. *If $(\perp_i, \boxdot) : i = 1,2,3$ is of type (1), $a \boxdot x = 1$ for all $a > 0$ and for all $x > 0$, or $a \boxdot x = g_3^{-1}(g_1(a) \cdot g_2(x))$ for all $a \in [0,1]$ and for all $x \in [0,1]$.*

2. *If* $(\perp_i, \boxdot) : i = 1,2,3$ *is of type (3), then* $a \boxdot x = 1$ *for all* $a > 0$ *and for all* $x > 0$
3. *Suppose that* $(\perp_i, \boxdot) : i = 1,2,3$ *is of type (4).*
   (a) *If* $\perp_1$ *is Archimedean, then* $a \boxdot x = 1 \boxdot x$ *for all* $a \in [0,1]$ *and for all* $x \in [0,1]$.
   (b) *If* $\perp_2$ *is Archimedean, then* $a \boxdot x = a \boxdot 1$ *for all* $a \in [0,1]$ *and for all* $x > 0$.
   (c) *If* $\perp_1$ *and* $\perp_2$ *are both Archimedean, then* $a \boxdot x = 1 \boxdot 1$ *for all* $a > 0$ *and for all* $x > 0$.

As seen in the previous proposition, only t-system of type (1) and (2) have rich structure. So we consider only Archimedean t-systems (type(1)) and t-systems of $\vee$ type (type(2)).

For a given t-conorm $\perp$, we define an operation $-_\perp$ by

$$a -_\perp b := \inf\{c | b \perp c \geq a\}$$

for all $(a,b) \in [0,1]^2$.

A $\perp-$ decomposable fuzzy measure m is called normal if $\perp = \vee$ or $g \circ m$ is an infinite additive measure or $g \circ m$ is a finite additive measure, where g is a generator of $\perp$.

Definition 12 Let $(X, 2^X, m)$ be a fuzzy measure space and $(\perp_i, \boxdot) : i = 1,2,3$ be a t-system. If m is a normal $\perp_2-$ decomposable fuzzy measure, then for a function $f : X \to [0,1]$ ($f = \perp_1{}_{i=1}^n a_i 1_{D_i}$ where $D_i \cap D_j \neq \emptyset$ for $i \neq j$) the t-conorm integral is defined as follows:

$$(T) \int f \boxdot dm := \perp_3{}_{i=1}^n a_i \boxdot m(D_i).$$

Example 5 Let $([0,1], S, U)$ be a conditionally distributive semiring defined in Example 1. In the case of $\perp_1 = \perp_2 = \perp_3 = S$ and $\boxdot = U$, the $t-$ conorm integral in Definition 12 coincides with $(S,U)-$ integral by Klement, Mesiar and Pap [9].

Definition 13 Let $(X, 2^X, m)$ be a fuzzy measure space and $(\perp_i, \boxdot) : i = 1,2,3$ be a t-system. For a function $f : X \to [0,1]$ ($f = \perp_1{}_{i=1}^n a_i 1_{D_i}$ where $D_i \cap D_j \neq \emptyset$ for $i \neq j$) the generalized t-conorm integral is defined as follows:

$$(GT) \int f \boxdot dm := \perp_3{}_{i=1}^n (a_i - a_{i-1}) \boxdot m(D_i).$$

If m is a normal $\perp_2$ decomposable fuzzy measure, the generalized t-conorm integral (Definition 13) coincides with the t-conorm integral (Definition 12).

Theorem 14 *Let* $(X, 2^X, m)$ *be a fuzzy measure space* , $(\perp_i, \boxdot) : i = 1,2,3$ *be a t-system and* $f : X \to [0,1]$ *be a measurable function.*

1. *If t-system is Archimedean (type (1)) and m is $\perp_2$ decomposable measure, the t-conorm integral is expressed as*

$$(T) \int f \boxdot dm = g_3^{-1}\left(\int g_1(f) d(g_2 \circ m)\right).$$

2. *If t-system is Archimedean (type (1)), the generalized t-conorm integral is expressed as*

$$(GT) \int f \boxdot dm = g_3^{-1}\left((C) \int g_1(f) d(g_2 \circ m)\right).$$

3. *If t-system is of $\vee$ type (type (2)), the generalized t-conorm integral is expressed as*

$$(GT) \int f \boxdot dm = \sup_{\alpha \in [0,1]} [\alpha \boxdot m(\{x | f(x) > \alpha\})].$$

Example 6 1. If $\perp_1 = \perp_2 = \perp_3 = \hat{+}$ and $\boxdot = \cdot$, the generalized t-conorm integral is Choquet integral.
2. If $\perp_1 = \perp_2 = \perp_3 = \vee$ and $\boxdot = \wedge$, the generalized t-conorm integral is Sugeno integral.

Since a generalized t-conorm integral for an Archimedean t-system is represented in terms of a Choquet integral, from a purely mathematical view, the only two essentially different integral with respect to a fuzzy measure are Choquet integral and the generalized t-conorm integral based on $(\vee, \vee, \vee, \boxdot)$ (Sugeno type integral). In particular, as a t-conorm integral for an Archimedean t-system is represented in terms of a classical integral, the classical integration theory is applicable in this case. This means that the classical statistic methods can be applied for data mining in the case that we use a t-conorm integral for Archimedean t-system.

4 Conditions for Choquet or Sugeno integral

In this section, we show the necessary and sufficient conditions for an aggregation functions to be the Choquet or Sugeno integral. The comonotonicity plays the central role in this section.

Definition 15 Let $\mathcal{I}$ be an aggregation function. We say:

1. $\mathcal{I}$ is comonotonic additive if and only if for comonotonic $x, y \in [0,1]^n$,

$$\mathcal{I}(x+y) = \mathcal{I}(x) + \mathcal{I}(y),$$

2. $\mathcal{I}$ is monotone if and only if for $x, y \in [0,1]^n$

$$x \leq y \Rightarrow \mathcal{I}(x) \leq \mathcal{I}(y),$$

3. $\mathcal{I}$ is comonotonic monotone if and only if for comonotonic $x, y \in [0,1]^n$
$$x \le y \Rightarrow \mathcal{I}(x) \le \mathcal{I}(y),$$
4. $\mathcal{I}$ is min homogeneous if and only if for $a \in [0,1]$
$$\mathcal{I}(a \wedge x) = a \wedge \mathcal{I}(x),$$
5. $\mathcal{I}$ is comonotonic maxitive if and only if for comonotonic $x, y \in [0,1]^n$
$$\mathcal{I}(x \vee y) = \mathcal{I}(x) \vee \mathcal{I}(y).$$

Suppose that the aggregation function $\mathcal{I}$ is comonotonic additive, then we have
$$\mathcal{I}(ax) = a\mathcal{I}(x)$$
for $a \in [0,1]$ and $x \in [0,1]^n$, that is, $\mathcal{I}$ is positive homogeneous [16].

As the conditions for an aggregation function to be the Choquet integral, We have the next theorem, which is a slightly generalized Schmeidler's theorem [19].

Theorem 16 *Let $\mathcal{I} : [0,1]^n \to R_+$ be comonotonic monotone and comonotonic additive functional with $\mathcal{I}(1_X) = 1$. There exists a fuzzy measure μ on $(X, 2^X)$ such that*
$$\mathcal{I}(x) = C_\mu(x)$$
for all $x \in [0,1]^n$.

Proof. Let $\mathcal{I} : [0,1]^n \to R_+$ be comonotonic monotone and comonotonic additive functional. For all $x \in [0,1]^n$, we have
$$x = \sum_{i=1}^{n} (x^{(i)} - x^{(i-1)}) 1_{\{(i)\cdots(n)\}},$$
where $1_{\{(i)\cdots(n)\}}$ is a characteristic function of $\{(i)\cdots(n)\}$, that is,
$$1_A(x) := \begin{cases} 1 & \text{if } x \in A \\ 0 & \text{if } x \notin A \end{cases}$$
for $A \in 2^X$.

Define the set function on 2^X by
$$\mu(A) := \mathcal{I}(1_A)$$
for $A \in 2^X$. If $A \subset B$, we have $\mu(A) \le \mu(B)$ since 1_A and 1_B are comonotonic and $1_A \le 1_B$. Then it follows from positive homogeneity and comonotonic

additivity that

$$\begin{aligned}
\mathcal{I}(x) &= \mathcal{I}(\sum_{i=1}^{n}(x^{(i)} - x^{(i-1)})1_{\{(i)\cdots(n)\}}) \\
&= \sum_{i=1}^{n}\mathcal{I}((x^{(i)} - x^{(i-1)})(1_{\{}(i)\cdots(n)\})) \\
&= \sum_{i=1}^{n}(x^{(i)} - x^{(i-1)})\mathcal{I}(1_{\{}(i)\cdots(n)\}) \\
&= \sum_{i=1}^{n}(x^{(i)} - x^{(i-1)})\mu((\{(i)\cdots(n)\})) = C_\mu(x).
\end{aligned}$$

for $x \in [0,1]^n$. □

Applying Theorem 16, we can show the next corollary.

Corollary 17 *Let $\mathcal{I}$ be an aggregation function. If $\mathcal{I}$ is monotone and additive on every $\mathcal{D}_\pi; \pi \in \mathcal{S}_n$, there exists a fuzzy measure μ on $(X, 2^X)$ such that*

$$\mathcal{I}(x) = C_\mu(x)$$

for all $x \in [0,1]^n$.

This corollary says that we should do the following to confirm whether the Choquet integral is applicable in data mining: (i) dividing $[0,1]^n$ into the comonotonic class $\mathcal{D}_\pi$ for $\pi \in \mathcal{S}_n$, (ii) regarding the aggregation $\mathcal{I}$ as linear on $\mathcal{D}_\pi$.

Next we show the conditions for an aggregation function to be the Sugeno integral. The next theorem is based on Ralescu and Sugeno [18], that is essentially the same as Marichal [12] and Benvenuti and Mesiar [2].

Theorem 18 *Let $\mathcal{I}$ be an aggregation functional with following properties:*

1. *$\mathcal{I}$ is comonotonic monotone,*
2. *$\mathcal{I}$ is comonotonic maxitive*
3. *$\mathcal{I}$ is min homogeneous.*
4. *$\mathcal{I}(1_X) = 1$. Then there exists a fuzzy measure $\mathcal{I}$ such that*

$$\mathcal{I}(x) = S_\mu(x)$$

Proof. Suppose that $A, B \in 2^X$. Then we have $1_A \leq 1_B$ and 1_A and 1_B are comonotonic, where 1_A is a characteristic function of A, that is, $1_A(x) = 1$ if $x \in A$ and $1_A(x) = 0$ if $x \notin A$. Therefore we can define a fuzzy measure μ by

$$\mu(A) := \mathcal{I}(1_A)$$

for $A \in 2^X$. Let $x \in [0,1]^n$. Since we have

$$x = \bigvee_{i=1}^{n} x^{(i)} \wedge 1_{\{(i)\cdots(n)\}},$$

it follows from *2* and *3* that

$$\begin{aligned}\mathcal{I}(x) &= \bigvee_{i=1}^{n} \mathcal{I}(x^{(i)} \wedge 1_{\{(i)\cdots(n)\}}) \\ &= \bigvee_{i=1}^{n} x^{(i)} \wedge \mathcal{I}(1_{\{(i)\cdots(n)\}}) \\ &= \bigvee_{i=1}^{n} x^{(i)} \wedge \mu(\{(i)\cdots(n)\}) = S_\mu(x). \square\end{aligned}$$

As the corollary, we have the next conditions for an aggregation function to be the Sugeno integral with respect to a possibility measure Π.

Corollary 19 *Let the aggregation functional $\mathcal{I}$ have the properties:*

1. $\mathcal{I}(x \vee y) = \mathcal{I}(x) \vee \mathcal{I}(y)$ *for every* $x, y \in [0,1]^n$,
2. $\mathcal{I}(a \wedge x) = a \wedge \mathcal{I}(x)$ *for* $a \in [0,1]$,
3. $\mathcal{I}(1_X) = 1$.

Then there exists a possibility measure Π such that

$$\mathcal{I}(x) = S_\Pi(x).$$

References

1. T. E. Armstrong, Chebyshev inequalities and comonotonicity, *Real Analysis Exchange,* 19(1), 1993/1994, 266-268.
2. P. Benvenuti and R. Mesiar, A note on Sugeno and Choquet integrals, Proc. 8th Intern. Conf. Information Processing and Management of Uncertainty in Knowledge-based Systems (IPMU 2000) (2000) 582-585.
3. G.Choquet . Theory of capacities. *Ann. Inst. Fourier, Grenoble.* 5 (1955) 131-295.
4. C. Dellacherie, Quelques commentaires sur les prolongements de capacités, *Séminaire de Probabilités 1969/1970, Strasbourg, Lecture Notes in Mathematics,* 191, Springer, 1971, 77- 81.
5. D. Denneberg, *Non additive measure and Integral,* Kluwer Academic Publishers, Dordrecht, 1994.
6. M. Grabisch, T. Murofushi, and M. Sugeno, eds., : Fuzzy Measures and Integrals: Theory and Applications, Physica-Verlag, Berlin, 2000.
7. M. Grabisch, H.T. Nguyen and E. A. Walker, *Fundamentals of uncertainty calculi with applications to fuzzy inference,* Kluwer Academic Publishers, Dordorecht 1995.

8. E.P. Klement, R. Mesiar and E. Pap, *Triangular Norms*, Kluwer Academic Publishers, Dordrecht, 2000.
9. E.P. Klement, R. Mesiar and E. Pap, Integration with respect to decomposable measures, based on a conditionally distributive semiring on the unit interval. *International Journal of Uncertainty, Fuzziness and Knowledge-Based Systems*,8 (2000), no. 6, 701–717 .
10. D.H. Krantz, R.D. Luce, P. Suppes and A. Tversky, *Foundation of Measurement,* Vol.I, Academic Press, New York, (1971).
11. C.H.Ling, Representation of associative functions, *Publ. Math. Debrecen*, 12 (1965),189-212.
12. J.L. Marichal, On Sugeno integral as an aggregation function, *Fuzzy sets and systems,* 114, (2000), 347-365.
13. R.Mesiar, Choquet-like integrals. *J. Math. Anal. Appl.* 194, No.2,(1995) 477-488.
14. T. Murofushi and M. Sugeno, "An interpretation of fuzzy measures and the Choquet integral as an integral with respect to a fuzzy measure," *Fuzzy Sets and Systems,* 29 , (1989), 201-227.
15. T. Murofushi and M. Sugeno, Fuzzy t-conorm integral with respect to fuzzy measures: generalization of Sugeno integral and Choquet integral, *Fuzzy Sets and Systems,* 42, no.1 (1991) ,57-71.
16. Y. Narukawa, T. Murofushi, and M. Sugeno: Regular fuzzy measure and Representation of comonotonically additive functionals , Fuzzy Sets and Systems, 112 (No 2) , (2000), 177-186.
17. E. Pap, An integral generated by a decomposable measure. *Zb. Rad., Prir.-Mat. Fak., Univ. Novom Sadu, Ser. Mat.* 20, No.1,(1990) 135-144.
18. D.A. Ralescu and M. Sugeno, Fuzzy integral representation, *Fuzzy sets and Systems*, 84,(1996) 129-133.
19. D. Schmeidler, "Integral representation without additivity," *Proceedings of the American Mathematical Society,* 97 , (1986), 253-261.
20. M. Sugeno, *Theory of fuzzy integrals and its applications,* Doctoral Thesis, Tokyo Institute of Technology, (1974).
21. M. Sugeno and T. Murofushi, Pseudo additive measures and integrals, *J. Math. Anal. Appl.* 122 (1987) 197-222.
22. V. Torra, Information Fusion in Data Mining, Chapter in this Volume.
23. S. Weber, $\perp$-decomposable measures and integrals for Archimedean t-conorms $\perp$. *J. Math. Anal. Appl.* 101, (1984),114-138.
24. S. Weber, Two integrals and some modified versions - critical remarks, *Fuzzy sets and Systems*, 20(1986)97-105.
25. R.R. Yager and A. Rybalov. Uninorm aggregation operators, *Fuzzy sets and Systems*,80 (1996), 111-120.
26. B. L. van der Waerden, *Mathematical statistics*, Springer, Berlin, 1969.

Data Mining Using a Probabilistic Weighted Ordered Weighted Average (PWOWA) Operator

H. B. Mitchell

Elta Electronics Industries Ltd.
Intelligent Centers Department (Section 3424)
Signal Processing and Computer Division
Ashdod, Israel
Email: mitchell@is.elta.co.il

Abstract. The Weighted Ordered Weighted Average (WOWA) operator is a powerful operator used for aggregating a set of M input arguments which may derive from different sources. The WOWA operator allows the user to take into account both the importance or reliability of the different information sources and the relative position of the argument values. In this chapter we describe a powerful new probabilistic weighted ordered weighted average (PWOWA) operator, which is simple and fast to implement, robust and whose parameters may all be given a direct physical interpretation. The chapter concludes with a demonstration of the new operator in a practical example involving lossless image compression.

1 Introduction

Data mining is concerned with the extraction of useful knowledge from databases containing large, and sometimes very large, amounts of (raw) data. One method for performing this task is to construct a data model which explains a particular variable z in terms of one or more other variables a_i. In mathematical terms, our aim is to find a function f such that $f(a_1, a_2, \ldots)$ is a good approximation to z:

$$z \simeq f(a_1, a_2, \ldots) \ . \tag{1}$$

One approach to constructing the function f is to use a parametric averaging operator F_{AVE} [1–7]. Such operators have a fixed algebraic form and one or more free parameters $\alpha, \beta, \ldots, \lambda$, which we may vary for optimum performance. We emphasize the fact that f has free parameters $\alpha, \beta \ldots, \lambda$, by rewriting (1) as

$$z \simeq f(a_1, a_2, \ldots, | \ \alpha, \beta, \ldots, \lambda) \ . \tag{2}$$

One point to note about the parametric average, and other aggregation functions, is that they all act in a deterministic manner: the same input values

a_i and the same free parameter values $\alpha, \beta, \ldots, \lambda$, always generate the same output value z.

In this chapter we describe a new type of averaging operator which acts in a probabilistic manner. In this operator the same input values a_i and the same free parameter values $\alpha, \beta, \ldots, \lambda$, do not necessarily generate the same output value z. This non-deterministic behaviour, makes the new operator very powerful, a demonstration of which is given in section 5.

The chapter is organized as follows. In section 2 we briefly describe the traditional parametric averaging operator. Then, in section 3 we describe in detail the weighted ordered weighted average (WOWA) operator since this forms the basis of the new probabilistic operator. In section 4 we describe the new operator and in section 5 we illustrate its use in a practical example involving lossless image compression. Finally, the chapter ends with a brief summary and conclusion in section 6.

2 Parametric Averaging Operators

Given a set of M input variables $a_i, i \in \{1, 2, \ldots, M\}$, which may, or may not, derive from different sources of information, we may identify two distinct types of averaging operators:

- Weighted average (WA) [2,8–10]. In this operator the free parameters $\alpha, \beta, \ldots, \lambda$ describe the importance or reliability of the different information sources. Thus, if p_i denotes the importance, or reliability, of the information source that generated the ith input variable a_i, then the weighted average is defined as

$$z = f_{\mathrm{WA}}\left(a_1, a_2, \ldots, a_M \mid p_1, p_2, \ldots, p_M\right), \\ = \sum_{i=1}^{M} p_i a_i , \tag{3}$$

  where

$$p_i \in [0, 1] , \tag{4}$$

$$\sum_{i=1}^{M} p_i = 1 . \tag{5}$$

- Ordered weighted average (OWA) [4,11–15]. In this operator, the free parameters $\alpha, \beta, \ldots, \lambda$, take into account the relative values of the different variables. Thus if w_i denote the relative importance of the input variable with the ith largest input value, then the ordered weighted average is defined as

$$z = f_{\mathrm{OWA}}(a_1, a_2, \ldots, a_M \mid w_1, w_2, \ldots, w_M) , \\ = \sum_{i=1}^{M} w_i a_{(i)} , \tag{6}$$

where $a_{(i)}$ is the value of the variable which has the ith largest value, and

$$w_i \in [0,1] , \tag{7}$$

$$\sum_{i=1}^{M} w_i = 1 . \tag{8}$$

2.1 Optimizing the free parameters

The parametric averaging operators have the free parameters $\alpha, \beta, \ldots, \lambda$, which may be adjusted for optimum performance. One method for doing this is to define some performance measure and then to vary the parameters until the optimum performance measure is obtained on a set of examples. For example, suppose we wish to find the source reliabilities p_i for the WA operator which give the minimum mean square error. Suppose our database contains N examples of the variable z and the input data variables a_i, where $z(k)$ and $a_i(k), k \in \{1, 2, \ldots, N\}$, are the corresponding z and a_i values in the ith example, then the optimum p_i values are p_i^*, where

$$(p_1^*, p_2^*, \ldots, p_M^*) = \min\Big(\sum_{k=1}^{N}\Big(z(k) - \sum_{i=1}^{M} p_i a_i(k)\Big)^2\Big) , \tag{9}$$

subject to the constraints

$$\sum_{i=1}^{M} p_i^* = 1 , \tag{10}$$

$$p_i^* \in [0,1] . \tag{11}$$

Sometimes we wish to find the optimum p_i by minimizing the maximum absolute error or the mean absolute error. In this case we would, respectively, replace (9) with (12) or with (13):

$$(p_1^*, p_2^*, \ldots, p_M^*) = \arg\min\Big(\max_k \Big\| z(k) - \sum_{i=1}^{M} p_i a_i(k) \Big\|\Big) , \tag{12}$$

$$(p_1^*, p_2^*, \ldots, p_M^*) = \arg\min\Big(\sum_{k=1}^{N} \Big\| z(k) - \sum_{i=1}^{M} p_i a_i(k) \Big\|\Big) . \tag{13}$$

The minimum mean square error, the minimum maximum absolute error and the minimum mean absolute error are three widely used traditional performance measures. However, in certain applications, other performance measures may be more appropriate and should be used. An example of this is given in section 5 where our performance measure is the minimum first-order entropy.

In general the optimum parameters must be found numerically. For this purpose dynamic programming, genetic algorithms and other optimization techniques have been used [16–22].

2.2 Outliers

In writing the variable z as a function $f(a_1, a_2, \ldots, a_M \mid \alpha, \beta, \ldots, \lambda)$ we have implicitly assumed that z depends on each, and every, input variable a_i. However, sometimes this is not true. When this happens, we refer to those input variables which have *no* influence, direct or indirect, as *outliers.*

In general the presence of a outlier x will not adversely affect the performance of a WA function f_{WA}. The reason is that in the WA function we may *effectively* exclude an outlier x by setting its weight, p_x, to zero.

This is not, however, the case for the OWA function f_{OWA}. Here a single outlier x may render the function f_{OWA} so inaccurate that it is simply not worthwhile to use an OWA function. The following example gives a demonstration of just such a situation.

Example 1. We consider the database D in Table 1 which contains N examples of the variable z, the outlier x and the input data variables a_1, a_2, a_3. The kth example contains the values $z(k), x(k)$ and $a_1(k), a_2(k), a_3(k)$.

The variable z is an OWA of the input variables a_1, a_2, a_3:

$$z(k) = 0.1a_{(i)}(k) + 0.7a_{(2)}(k) + 0.2a_{(3)}(k), k \in \{1, 2, \ldots, N\} , \tag{14}$$

where

$$a_{(i)}(k) = a_j(k) \text{ if } a_j(k) \text{ is the } i \text{ th largest } a(k) \text{ value} . \tag{15}$$

We now try to approximate z with an OWA of *all* the input variables, including the outlier x. Let $z'(k), k \in \{1, 2, \ldots, N\}$, denote the corresponding approximate $z(k)$ values, then

$$z'(k) = \sum_{i=1}^{4} w'_i a'_{(i)}(k) , \tag{16}$$

where

$$a'_i(k) = \begin{cases} a_i(k) \text{ if } i \in \{1, 2, 3\} , \\ x(k) \ \text{ if } i = 4 . \end{cases} \tag{17}$$

In Table 2 we give the values of $z(k)$, $z'(k)$ and the errors $e(k) = (z(k) - z'(k))$. The weights used in the OWA function were optimized for minimum mean square error. We see how a single outlier may cause substantial errors in the OWA function: the mean square error in Table 2 is approximately 60.

The reason why the OWA function is sensitive to the presence of outliers is as follows:

In using the OWA function we assume we may re-order the input variables in a *consistent* manner. The following example shows how a single outlier may prevent the consistent re-ordering of a set of input variables.

Table 1. The database D

k	z	a_1	a_2	a_3	x
1	34.00	33	29	40	3
2	33.90	29	22	57	87
3	50.70	25	59	52	13
4	30.30	0	33	36	99
5	88.90	94	43	94	50
6	93.80	97	96	72	65
7	43.40	86	3	37	9
8	39.60	33	1	82	3
9	43.30	42	3	68	33
10	42.10	56	42	15	6

Table 2. Values of z, z' and e calculated using an OWA function

k	z	z'	e
1	34.00	30.71	3.29
2	33.90	54.90	-21.04
3	50.70	48.85	1.85
4	30.30	35.07	-4.77
5	88.90	89.54	-0.64
6	93.80	93.33	0.47
7	43.40	35.66	7.74
8	39.60	31.81	7.79
9	43.30	39.57	3.73
10	42.10	39.33	2.77

The weight vector used is $\mathbf{w}' = [\,0.083\ 0.005\ 0.879\ 0.033\,]$

Example 2. Consider $(M+1)$ input variables $a_1, a_2, \ldots, a_M, x$ in which the a_i are true data variables and x is an outlier. For simplicity, we assume that

$$a_i = i\ . \tag{18}$$

Since x is an outlier it may assume any random value. We consider two extreme cases:

- In the first case, $x < 1$. In this case, the re-ordered input variables are

$$a'_{(i)} = \begin{cases} x & \text{if } i = 1 \,, \\ a_{i-1} & \text{if } i \in \{2,3,\ldots,(M+1)\} \,. \end{cases} \tag{19}$$

- In the second case, $x > M$. In this case, the re-ordered variables are

$$a''_{(i)} = \begin{cases} a_i & \text{if } i \in \{1,2,\ldots,M\} \,, \\ x & \text{if } i = (M+1) \,. \end{cases} \tag{20}$$

Comparing $a'_{(i)}$ and $a''_{(i)}$ we see how the presence of an outlier may prevent us from consistently re-ordering a set of input variables.

3 Weighted Ordered Weighted Average Function

Given the input variables $a_1, a_2, \ldots, a_M$, let $a_{(i)}$ denote the value of the input variable which has the ith largest value and $p_{(i)}$ the corresponding reliability value. In mathematical terms,

$$\left.\begin{aligned} a_{(i)} &= a_j \\ p_{(i)} &= p_j \end{aligned}\right\} \text{if } a_j \text{ is the } i\text{th largest input variable} \,. \tag{21}$$

Torra [23,24] defines the weighted ordered weighted average (WOWA) operator as:

$$F_{\text{WOWA}} = \sum_{i=1}^{M} \Omega_i a_{(i)} \,, \tag{22}$$

where

$$\Omega_i = W\left(\sum_{j=1}^{i} p_{(i)}\right) - W\left(\sum_{j=1}^{i-1} p_{(i)}\right) , \tag{23}$$

and W is a monotonic increasing function that interpolates the points $(0,0)$, $(1/M, w_1), (2/M, (w_1+w_2)) \ldots, (i/M, \sum_{j=1}^{i} w_j), \ldots, (1,1)$. W is required to be a straight line when the points can be interpolated in this way [23–26].

Torra [27–30] and others [31,32] have shown the versatility of the WOWA operator in several applications. We suggest that one reason why the WOWA operator has proved so useful, especially in applications involving in data mining, is that the WOWA behaves like an OWA operator but without its extreme sensitivity to outliers. This point is brought out very clearly in the following example.

Example 3. We use the same database D given in Table 1 and as before we let

$$a'_i(k) = \begin{cases} a_i(k) & \text{if } i \in \{1,2,3\} \,, \\ x(k) & \text{if } i = 4 \,. \end{cases} \tag{24}$$

This time we use a WOWA function to approximate the $z(k), k \in \{1, 2 \ldots, N\}$. If $z'(k)$ is the corresponding approximate $z(k)$ value, then

$$z'(k) = \sum_{i=1}^{4} \Omega'_i a'_{(i)}(k) \ , \tag{25}$$

where the weights Ω'_i depend on the weight vectors $\mathbf{w}' = [w'_1\ w'_2\ w'_3\ w'_4]$ and $\mathbf{p}' = [p'_1\ p'_2\ p'_3\ p'_4]$ through (23). In table 3 we give the values of $z(k)$, $z'(k)$

Table 3. Values of z, z' and e calculated using a WOWA function

k	z	z'	e
1	34.00	34.74	-0.74
2	33.90	34.60	-0.70
3	50.70	43.91	6.79
4	30.30	20.46	9.84
5	88.90	85.65	3.25
6	93.80	89.16	4.64
7	43.40	42.31	1.09
8	39.60	44.27	-4.67
9	43.30	42.75	0.55
10	42.10	39.55	2.55

The weight vectors used are $\mathbf{w}' = [0.3\ 0.3\ 0.15\ 0.25]$, $\mathbf{p}' = [0.33\ 0.33\ 0.33\ 0]$

and the errors $e(k) = (z(k) - z'(k))$. The weight vectors $\mathbf{w}'$ and $\mathbf{p}'$ used in Table 3 were optimized for minimum mean square error. We see how the use of the WOWA function has enabled us to substantially reduce the size of the errors $e(k)$ as compared to those obtained with an OWA function. The mean square error obtained with the WOWA function is approximately 20.

4 Probabilistic Weighted Ordered Weighted Average Operator

Although the WOWA operator is versatile it requires the user to define an appropriate interpolation function W. This is not a simple task and it may be computationally difficult. Furthermore it is difficult to give a direct physical interpretation to the different possible interpolation functions.

In this chapter we introduce a probabilistic weighted ordered weighted average (PWOWA) which is both simple to implement and may be given a direct physical interpretation. Suppose we let the degree of importance p_i be the probability of appearance of the variables $a_i, i \in \{1,2,\ldots,M\}$. Then given the M arguments $a_1, a_2, \ldots, a_M$ and their importances $p_1, p_2, \ldots, p_M$, we may define a new set of variables $b_1, b_2, \ldots, b_M$ using (26):

$$b_i = \begin{cases} a_1 & \text{if } 0 \le r_i < p_1 \ , \\ a_2 & \text{if } p_1 \le r_i < (p_1 + p_2) \ , \\ \vdots & \\ a_M & \text{if } (p_1 + p_2 + \cdots + p_{M-1}) \le r_i \le 1 \ , \end{cases} \tag{26}$$

where $r_1, r_2, \ldots, r_M$ are a set of random numbers chosen uniformly between 0 and 1.

It is straightforward to see that (26) is consistent with our interpretation of p_i as the probability of appearance of the variable a_i: Suppose we repeat (26) L times in a monte carlo simulation. Let $b_{i|l}, i \in \{1,2,\ldots,M\}$, denote the set of new variables in the lth monte carlo run. Then as L increases, the mean number of variables $b_{1|l}, b_{2|l}, \ldots, b_{M|l}, l \in \{1,2,\ldots,L\}$, which have a value equal to a_i will tend asymptotically to p_i. Mathematically, we have

$$\frac{1}{ML}\sum_{l=1}^{L}\sum_{j=1}^{M} \epsilon_{j|l} \rightarrow p_i \ , \tag{27}$$

where

$$\epsilon_{j|l} = \begin{cases} 1 \text{ if } b_{j|l} = a_i \ , \\ 0 \text{ otherwise} \ . \end{cases} \tag{28}$$

In the new PWOWA operator we obtain a single output variable z by aggregating the variables $b_{i|l}$ obtained in a monte carlo simulation. There are two different ways of performing this aggregation:

- In the first approach we aggregate the $b_{1|l}, b_{2|l}, \ldots, b_{M|l}$ for each l, using the standard OWA operator. We then aggregate over l using any parametric average operator F_{AVE}. Mathematically this process defines a PWOWA operator:

$$F_{\text{PWOWA}} = F_{\text{AVE}}(Q_1, Q_2, \ldots, Q_L) \ , \tag{29}$$

where

$$Q_l = \sum_{i=1}^{M} w_i b_{(i)|l} \ , \tag{30}$$

and $b_{(i)|l}$ is the ith largest value amongst $(b_{1|l}, b_{2|l}, \ldots, b_{M|l})$.
- In the second approach we aggregate the *re-ordered* variables $b_{(i)|1}, b_{(i)|2}, \ldots, b_{(i)|L}$ for each i, using any parametric average operator F_{AVE}. We then aggregate over i using the standard OWA operator. Mathematically the process defines a second type of PWOWA operator:

$$F_{\text{PWOWA}} = \sum_{i=1}^{M} w_i Q_{(i)} \,, \tag{31}$$

where

$$Q_{(i)} = F_{\text{AVE}}(b_{(i)|1}, b_{(i)|2}, \ldots, b_{(i)|L}) \,. \tag{32}$$

There are many different possible choices for F_{AVE}. Two simple choices are the arithmetic mean and the median operator. Using these operators in (29) and (31) we obtain four PWOWA operators

- $$F^{(1)}_{\text{PWOWA}} = \frac{1}{L}\sum_{l=1}^{L}\Big(\sum_{i=1}^{M} w_i b_{(i)|l}\Big) \,, \tag{33}$$
- $$F^{(2)}_{\text{PWOWA}} = \text{median}\Big(\sum_{i=1}^{M} w_i b_{(i)|1}, \sum_{i=1}^{M} w_i b_{(i)|2}, \ldots, \sum_{i=1}^{M} w_i b_{(i)|L}\Big) \,, \tag{34}$$
- $$F^{(3)}_{\text{PWOWA}} = \sum_{i=1}^{M} w_i \Big(\frac{1}{L}\sum_{l=1}^{L} b_{(i)|l}\Big) \,, \tag{35}$$
- $$F^{(4)}_{\text{PWOWA}} = \sum_{i=1}^{M} w_i \Big(\text{median}(b_{(i)|1}, b_{(i)|2}, \ldots, b_{(i)|L})\Big) \,. \tag{36}$$

In fact, however, only three of the four PWOWA operators are independent, since $F^{(3)}_{\text{PWOWA}} \equiv F^{(1)}_{\text{PWOWA}}$.

Proposition $F^{(3)}_{\text{PWOWA}} \equiv F^{(1)}_{\text{PWOWA}}$

Proof

$$
\begin{aligned}
F^{(3)}_{\mathrm{PWOWA}} &= \sum_{i=1}^{M} w_i \left(\frac{1}{L} \sum_{l=1}^{L} b_{(i)|l} \right), \\
&= \frac{w_1}{L} \left(b_{(1)|1} + b_{(1)|2} + \ldots + b_{(1)|L} \right) \\
&\quad + \frac{w_2}{L} \left(b_{(2)|1} + b_{(2)|2} + \ldots + b_{(2)|L}) \right) \\
&\quad \vdots \\
&\quad + \frac{w_L}{L} \left(b_{(M)|1} + b_{(M)|2} + \ldots + b_{(M)|L} \right), \\
&= \frac{1}{L} \left(w_1 b_{(1)|1} + w_2 b_{(2)|1} + \ldots + w_M b_{(M)|1} \right) \\
&\quad + \frac{1}{L} \left(w_1 b_{(1)|2} + w_2 b_{(2)|2} + \ldots + w_M b_{(M)|2}) \right) \\
&\quad \vdots \\
&\quad + \frac{1}{L} \left(w_1 b_{(1)|L} + w_2 b_{(2)|L} + \ldots + w_M b_{(M)|L} \right), \\
&= \frac{1}{L} \sum_{l=1}^{L} \left(\sum_{i=1}^{M} w_i b_{(i)|l} \right) \\
&= F^{(1)}_{\mathrm{PWOWA}} .
\end{aligned}
$$

Eqns. 33 - 36 define four PWOWA operators. The steps required to compute these new operators are:

1. Perform a monte carlo simulation of L runs.
2. For each monte carlo run create a new set of arguments $b_{i|l}$ using (26).
3. Aggregate the $b_{i|l}$ using, as appropriate, (33)- (36).

Example 4. We use the same database D given in Table 1 and as before we let

$$
a'_i(k) = \begin{cases} a_i(k) \text{ if } i \in \{1,2,3\} , \\ x(k) \;\; \text{if } i = 4 . \end{cases} \tag{37}
$$

This time we use each of the PWOWA operators in turn to approximate the $z(k)$. For each operator we separately optimized the weights $w_i^{(1)-(4)}$ and $p_i^{(1)-(4)}$ for minimum mean square error (see table 4). In table 5 we give the values of $z(k)$, the approximations $z'^{(1)}(k) - z'^{(4)}(k)$ and the errors $e^{(1)}(k)$- $- e^{(4)}(k)$. We see from Table 5 that by using the new operator we are able to very closely approximate the true z values: the mean square errors obtained with the four PWOWA operators are, respectively, 11.47, 1.15, 11.47 and 0.0.

Table 4. Optimum weights $w_i^{(1)-(4)}$ and $p_i^{(1)-(4)}$ for the new PWOWA operators

i	$w_i^{(1)}$	$p_i^{(1)}$	$w_i^{(2)}$	$p_i^{(2)}$	$w_i^{(3)}$	$p_i^{(3)}$	$w_i^{(4)}$	$p_i^{(4)}$
1	0.00	0.33	0.10	0.33	0.00	0.33	0.10	0.33
2	0.35	0.33	0.25	0.33	0.35	0.33	0.00	0.33
3	0.65	0.33	0.50	0.33	0.65	0.33	0.70	0.33
4	0.00	0.00	0.15	0.00	0.00	0.00	0.20	0.00

Table 5. Values of z, $z'^{(1)-(4)}$ and $e^{(1)-(4)}$ calculated using the new PWOWA operators ($L = 10$)

k	z	$z'^{(1)}$	$e^{(1)}$	$z'^{(2)}$	$e^{(2)}$	$z'^{(3)}$	$e^{(3)}$	$z'^{(4)}$	$e^{(4)}$
1	34.00	34.83	-0.83	34.05	-0.05	34.83	-0.83	34.00	0.00
2	33.90	38.11	-4.21	33.20	0.70	38.11	-4.21	33.90	0.00
3	50.70	48.76	1.94	50.35	0.35	48.76	1.94	50.70	0.00
4	30.30	27.57	2.73	30.15	0.15	27.57	2.73	30.30	0.00
5	88.90	84.00	4.90	88.90	0.00	84.00	4.90	88.90	0.00
6	93.80	89.67	4.13	93.60	0.20	89.67	4.13	93.80	0.00
7	43.40	44.61	-1.21	40.95	2.45	44.61	-1.21	43.40	0.00
8	39.60	45.07	-5.47	40.35	-0.75	45.07	-5.47	39.60	0.00
9	43.30	44.09	-0.79	45.25	-1.95	44.09	-0.79	43.30	0.00
10	42.10	38.64	3.46	41.40	0.70	38.64	3.46	42.10	0.00

5 Lossless image Compression

We illustrate the process of data modeling using an unusual example involving lossless image compression. In this example, our database is an input image and the raw data are the pixel gray-levels.

Many modern lossless image compression algorithms [33–36], including various recently proposed algorithms [37–42] and the new JPEG-LS international standard [43,44], work as follows. We scan the input image from top to bottom. Then, in each row we scan the pixel gray-levels from left to right. At any pixel (u, v), we predict its gray-level $g(u, v)$ by aggregating several basic predictors $a_i(u, v)$ (see Table 6). Each basic predictor may be either a previously scanned pixel gray-level (e. g. $a_1(u, v)$, $a_2(u, v)$, $a_4(u, v)$, $a_6(u, v)$) or a linear combination of several previously scanned pixel gray-levels (e. g. $a_3(u, v)$, $a_5(u, v)$, $a_7(u, v)$). The predicted gray-level for the pixel (u, v) is

$$h(u, v) = F_{\text{AVE}}\Big(a_1(u, v), a_2(u, v), \ldots, a_M(u, v)\Big) . \tag{38}$$

Table 6. Basic Predictors

Symbol	Predictor
$a_1(u,v)$	$N \equiv g(u-1,v)$
$a_2(u,v)$	$W \equiv g(u,v-1)$
$a_3(u,v)$	$N+W-NW$
$a_4(u,v)$	$NE \equiv g(u-1,v+1)$
$a_5(u,v)$	$(N+W)/2$
$a_6(u,v)$	$NW \equiv g(u-1,v-1)$
$a_7(u,v)$	$(NE+N)/2$

Instead of encoding the gray-level $g(u,v)$ we encode the *difference* $(g(u,v)-h(u,v))$. To a good approximation [33], the resultant bit-rate is the first-order entropy S:

$$S=\sum_d p(d)\log_2(p(d)) \ , \tag{39}$$

where $p(d)$ is the probability of appearance of the gray-level difference d. If the predictions are accurate, the differences will be very close to zero which we may efficiently encode using standard entropy codes. The first-order entropy is thus a good measure of how well the prediction function is performing.

In a series of detailed experiments Memon, Sippy and Wu [44] found that a simple OWA aggregation of the basic predictors $a_1(u,v), a_2(u,v)$ and $a_3(u,v)$ gave the best overall performance, i. e. the lowest first-order entropy S:

$$h(u,v)=\sum_{i=1}^{3} w_i a_{(i)}(u,v) \ , \tag{40}$$

where $\mathbf{w} \equiv [w_1 \ w_2 \ w_3] = [0 \ 1 \ 0]$. This predictor was later recommended by the Joint Programmers Expert Group (JPEG) for incorporation in the new JPEG-LS lossless image compression international standard [43,34]. Hereinafter, we shall denote the predictor as $h_{\text{JPEG-LS}}(u,v)$

Apart from the three basic predictors $a_1(u,v), a_2(u,v), a_3(u,v)$ many other predictors have been used in the past. Some of the more common ones are listed in Table 6.

We tested each of the four PWOWA operators defined in (33)-(36) by using them to aggregate all seven predictors $a_i(u,v), i \in \{1,2,\ldots,7\}$. We found, through an exhaustive search, that the best overall performance was obtained with $F^{(2)}_{\text{PWOWA}}$ using the set of p_i and w_i values given in table 7. Substituting the optimum p_i and w_i values into (38) we find the new predictor is given by

$$h_{\text{NEW}}(u,v)=\text{median}\left(b_{(4)|1}(u,v), b_{(4)|2}(u,v),\ldots,b_{(4)|L}(u,v)\right) \tag{41}$$

where

$$b_{i|l}(u,v) = \begin{cases} N & \text{if } 0 \leq r_{i|l} < 0.222\,, \\ (N+W-NW)/2 & \text{if } 0.222 \leq r_{i|l} < 0.555\,, \\ (N+W)/2 & \text{if } 0.555 \leq r_{i|l} < 0.777\,, \\ (NE+N)/2 & \text{if } 0.777 \leq r_{i|l} \leq 1\,, \end{cases} \tag{42}$$

$N \equiv g(u-1,v)$, $W \equiv g(u,v-1)$, $NE \equiv g(u-1,v+1)$ and $NW \equiv g(u-1,v-1)$ are the gray-levels of the four pixels which are adjacent to the current pixel (u,v) and which lie, respectively, north, west, north-east and north-west of (u,v), and $r_{i|l}, i \in \{1,2,\ldots,7\}, l \in \{1,2,\ldots,L\}$ are a set of random numbers chosen uniformly between 0 and 1. Regarding L we found the results were relatively insensitive to the actual value used. In order to keep the computational complexity low we, therefore, used $L = 10$.

Table 7. Optimum p_i and w_i values for h_{NEW} predictor

i	p_i	w_i
1	0.222	0
2	0.000	0
3	0.333	0
4	0.000	1
5	0.222	0
6	0.000	0
7	0.222	0

We compared the performance of new PWOWA predictor h_{NEW} and the JPEG-LS predictor $h_{\text{JPEG-LS}}$ by using them separately to encode a set of 14 monochromatic test pictures (Fig. 1). In table 8 we give the corresponding entropy values S_{NEW} and $S_{\text{JPEG-LS}}$ and the entropy differences $\Delta S = (S_{\text{NEW}} - S_{\text{JPEG-LS}})$. We see that for each, and every one, of the 14 test pictures the entropy difference is negative, showing that the new predictor consistently outperforms the JPEG-LS predictor: the average reduction in entropy is approximately 0.25 bits/pixel.

6 Conclusion

We have described a new type of weighted ordered weighted average operator which acts in a probabilistic manner. The new operator, known as the probabilistic weighted ordered weighted average (PWOWA) operator, is a powerful aggregation operator which is simple and straightforward to implement and its parameters may be given a direct physical interpretation.

Fig. 1. The 14 test pictures. Reading from top to bottom and in each row from left to right, the pictures are: Baloon, Barb, Board; Boat, Goldhill, Hotel; Zelda, Camera, Couple; Airplane, Tree, Peppers; Girl, House

We demonstrated the power of the new aggregation operator in a real-life application involving lossless image compression.

Acknowledgments

We gratefully acknowledge the many useful discussions held with P. A. Schaefer (Elta).

Table 8. Entropies S_{NEW} and $S_{\mathrm{JPEG-LS}}$ and the entropy differences ΔS

Picture	S_{NEW}	$S_{\mathrm{JPEG-LS}}$	ΔS
Baloon	3.1235	3.3751	-0.2516
Barb	5.2717	5.5178	-0.2461
Board	3.9925	4.2915	-0.2990
Boat	4.3738	4.6367	-0.2629
Goldhill	4.7432	4.9534	-0.2102
Hotel	4.7817	5.0706	-0.2889
Zelda	4.0940	4.3084	-0.2144
Camera	4.7734	5.0212	-0.2478
Couple	4.0751	4.4099	-0.3348
Airplane	5.2048	5.2469	-0.0421
Tree	5.3221	5.6096	-0.2875
Peppers	5.2834	5.4647	-0.1813
Girl	4.3809	4.6034	-0.2225
House	4.2150	4.5430	-0.3280

The pictures are widely available via the internet. For an up-to-date list of websites see e. g. http : //links.geocities.com/SiliconValley/Bay/1995/links.html

References

1. Torra V. (2002) Information fusion in data mining: outline pre-print
2. Torra V. (1999) On some relationships between hierarchies of quasiarithmetic means and neural networks. International J Intelligent Systems **14**: 1089–1098
3. Yager R. R. (1993) Toward a general theory of information aggregation. Information Sciences **68**: 191–206
4. Yager R.R., Kacprzyk J. (Editors) (1997) The ordered weighted averaging operators. Published by Kluwer Academic Press, Mass
5. Dubois D., Prade H. (1985). A review of fuzzy sets aggregation connectives. Information Sciences **36**: 85–121
6. Beliakov G., Warren J. (2001) Appropriate choice of aggregation operators in fuzzy decision support systems. IEEE Trans Fuzzy Systems **9**: 773–784
7. Kaymak U., Lemke, H. R. van Nauta (1994) Selecting an aggregation operator for fuzzy decision making. In: Proceedings The Third IEEE Conference on Fuzzy Systems, June 26–29, Orlando, Florida, **2**: 1418-1422.
8. Dubois D., Koning J. L. (1991) Social choice axioms for fuzzy set aggregation. Fuzzy Sets and Systems **43**: 257–274
9. Kaymak U., Lemke H. R. van Nauta (1993) A parametric generalized goal function for fuzzy decision making with unequally weighted objectives. In: Proceedings The Second IEEE Conference on Fuzzy Systems, March 28-April 1,**2**:1156–1160

10. Dyckhoff H., Pedrycz W. (1984) Generalized means as model of compensative connectives. Fuzzy Sets and Systems **14**: 143–154
11. Yager R. R. (1988) On ordered weighted averaging operators in multicriteria decision-making. IEEE Trans Systems Man and Cybernetics, **18**: 183–190
12. Yager R. R. (1992) Applications and extensions of OWA aggregations. International J Man-Machine Studies **37**: 103–132
13. Yager R. R. (1993) Families of OWA operators. Fuzzy Sets and Systems **59**: 125–148
14. Mitchell H. B., Estrakh D. D. (1998) An OWA operator with fuzzy ranks. International J. Intelligent Systems **13**: 69–81
15. Schaefer P. A. and Mitchell H. B. (1999) A generalized OWA operator. International J. Intelligent Systems **14**: 123–144
16. Torra V. (2000) Learning weights for weighted OWA operators. In: Proceedings IEEE International Conference Industrial Electronics, Control and Instrumentation, 22-28 Oct., Nagoya, Japan
17. Nettleton, D. F. and Torra, V. (2001) A comparison of active set method and genetic algorithm approaches for learning weighting vectors in some aggregation operators. International J Intelligent Systems **16**: 1069–1083
18. Filev, F. and Yager, R. R. (1994) Learning OWA weights from data. In: Proceedings The Third IEEE Conference on Fuzzy Systems, June 26–29, Orlando, Florida, **1**: 468–473
19. Filev D. P., Yager R. R. (1994) On the issue of obtaining the OWA operator weights. Fuzzy Sets and Systems. **94**: 157–169
20. Fuller R.,Majlender P. (2001) An analytic approach for obtaining maximal OWA operator weights. Fuzzy Sets and Systems **124**: 53–57
21. Yager R. R. (1995) Solving mathematical programming problems with OWA operators as objective functions. In: Proceedings 1995 IEEE International Conference on Fuzzy Systems, March 20–24, Yokohama, Japan, **3**:1441–1446
22. Torra V. (1999) On the learning of weights in some aggregation operators: the weighted mean and OWA operators. Mathware and Soft Computing **6**: 249–265
23. Torra V. (1997) The weighted OWA operator. International J Intelligent Systems **12**: 153–166
24. Torra V. (2000) The WOWA operator and the interpolation function w*: Chen and Otto's interpolation method revisited. Fuzzy Sets and Systems **113**: 389–396
25. Beliakov G. (2001) Shape preserving splines in constructing WOWA operators. Fuzzy Sets and Systems **121**: 549–550
26. Torra V. (2001) Authors Reply [Shape preserving splines in constructing WOWA operators]. Fuzzy sets and Systems **121**: 551
27. Torra V. (1996) Weighted OWA operators for synthesis of information In: Proceedings The Fifth IEEE International Conference on Fuzzy Systems, Sept 8–11, New Orleans, **2**:966-971
28. Torra V., Godo L. (1997) On defuzzification with continuous WOWA operators In: Proceedings Estylf '97, Tarragona, Spain, 227–232
29. Torra V. (2001) Sensitivity analysis for WOWA, OWA and WM operators. In: Proceedings 2001 IEEE international Symposium on Industrial Electronics, 12-16 June 2001, Pusan, South Korea. **1**: 134–137
30. Torra i Reventos v. (1999) Interpreting membership functions: A constructive approach. International J. Approximate Reasoning **20**: 191–207

31. Nettleton, D. F. and Hernandez, L. (1999) Evaluating reliability and relevance for WOWA aggregation of sleep apnea case data. In: Proceedings 1999 Eusflat-Estylf Joint Conference, Pulma de Mallorca, Spain
32. Nettleton D., Muniz J. (2001)Processing and representation of meta-data for sleep apnea diagnosis with an artificial intelligence approach. International J. Medical Informatics **63**: 77-89
33. Memon N. D., Sayood K. (1995) Lossless image compression: a comparative study. Proceeding SPIE **2418**: 8–20
34. Memon N. D., Wu X. L. (1997) Recent developments in context-based predictive techniques for lossless image compresion. Computer J. **40**: 127–136
35. Arps, R. B. and Truong, T. K. (1994) Comparison of international standards for lossless still image compression Proceedings IEEE **82**: 889–899
36. Savakis A. E. (2002) Evaluation of algorithms for lossless compression of continuous-tone images. J Electronic Imaging **11**: 75–86
37. Wu X. (1997). Lossless compression of continuous tone images via context selection, quantization and coding. IEEE Trans Image Process **5**: 1303–1310
38. Estrakh. D. D., Mitchell H. B., Schaefer P. A., Mann Y. Peretz Y. (2001) "Soft" median adaptive predictor for lossless picture compression. Signal Process **81**: 1985–1989
39. Mitchell H. B., Estrakh D.D. (1997) A modified OWA operator and its use in lossless DPCM image compression. International J. Uncertainty, Fuzziness and Knowledge-Based Systems **5**: 429–436
40. Mitchell H. B., Schaefer P. A. (2000) Multiple priorities in an induced ordered weighted averaging operator. International J. Intelligent Systems **15**: 317–327
41. Li X., Orchard M. T. (2001). Edge-directed prediction for lossless compression of natural images. IEEE Trans Image Processing **10**: 813–817
42. Deng G., Ye, H., Cahill, L. W. (2000) Adaptive combination of linear predictors for lossless image compression. IEE Proceedings Science, Measurement and Technology **147**: 414–419
43. Weinberger M. J., Seroissi G., Sapiro G. (2001). The LOCO-I Lossless image compression algorithm: principles and standardization into JPEG-LS. IEEE Trans Image Processing **9**: 1309–1324
44. Memon N., Sippy V., Wu X. (1996) A comparison of prediction schemes proposed for a new lossless image compression standard. In: Proceedings IEEE International Symposium on Circuits and Systems, Atlanta, USA, **3**: 309–312

Part 2:

Preprocessing Data

Mining Interesting Patterns in Multiple Data Sources

Ning Zhong

Department of Information Engineering
Maebashi Institute of Technology
460-1, Kamisadori-Cho, Maebashi 371-0816, Japan
E-mail: zhong@maebashi-it.ac.jp

Abstract. Since data are rarely specially collected/stored in a database for the purpose of mining knowledge in most organizations, a database always contains a lot of data that are redundant and not necessary for mining interesting patterns, as well as some interesting patterns may hide in multiple data sources rather than a single database. Hence peculiarity oriented multi-database mining are required. In the paper, *peculiarity rules* are introduced as a new class of patterns, which can be discovered from a relatively low number of peculiar data by searching the relevance among the peculiar data, as well as how to mine more interesting peculiarity rules in multiple data sources is investigated.

1 Introduction

Recently, it has been recognized in the data mining community that *multi-database mining* (MDM for short) is an important research topic [18,26]. Multi-database mining is to mine knowledge in multiple related data sources. It is a technique of information fusion in data mining to improve the performance of mining results. Generally speaking, the task of multi-database mining can be divided into three levels:

1. Mining from multiple relations in a database.
   Although theoretically, any relational database with multiple relations can be transformed into a single universal relation, practically this can lead to many issues such as universal relations of unmanageable sizes, infiltration of uninteresting attributes, loss of useful relation names, an unnecessary join operation, and inconveniences for distributed processing.
2. Mining from multiple relational databases.
   Some interesting patterns (concepts, regularities, causal relationships, and rules) cannot be discovered if we just search a single database because such patterns hide in multiple data sources basically [26]. In other words, data are rarely specially collected/stored in a database for the purpose of mining knowledge in most organizations.
3. Mining from multiple mixed-media databases.
   Many datasets in the real world contain more than just a single type of data [16,26]. For example, medical datasets often contain numeric

data (e.g. test results), images (e.g. X-rays), nominal data (e.g. person smokes/does not smoke), and acoustic data (e.g. the recording of a doctor's voice). How to handle such multiple data sources is a new, challenging research issue.

In order to discover *new, surprising, interesting* patterns hidden in data, peculiarity oriented multi-database mining are required. In the paper, *peculiarity rules* are introduced as a new class of patterns, which can be discovered from a relatively low number of peculiar data by searching the relevance among the peculiar data, as well as how to mine more interesting peculiarity rules in multiple data sources is investigated.

The rest of this paper is organized as follows: Section 2 gives a survey on related work in multi-database mining. Section 3 introduces Granular Computing for semantic heterogeneity among multiple data sources. Section 4 discusses interestingness and peculiarity. Section 5 presents a method of peculiarity oriented mining. Section 6 extends the peculiarity oriented mining into multi-database mining. Section 7 discusses a result of mining from the amino-acid data set. Finally, Section 8 gives concluding remarks.

2 Related Work on MDM

Multi-database mining (MDM for short) involves many related topics including interestingness checking, relevance, database reverse engineering, granular computing, and distributed data mining. Liu et al. proposed an interesting method for relevance measure and an efficient implementation for identifying relevant databases as the first step for multi-database mining [10]. Ribeiro et al. described a way for extending the INLEN system, which is a multi-strategy discovery system for mining knowledge from databases, for multi-database mining by the incorporation of primary and foreign keys as well as the development and processing of knowledge segments [13]. Wrobel extended the concept of foreign keys into foreign links because multi-database mining is also interested in getting to non-key attributes [18]. Aronis et al. introduced a system called WoRLD that uses spreading activation to enable inductive learning from multiple tables in multiple databases spread across the network [2]. Zhong et al. proposed a way for peculiarity oriented multi-database mining [26].

Database reverse engineering is a research topic that is closely related to multi-database mining. The objective of database reverse engineering is to obtain the domain semantics of legacy databases in order to provide meaning of their executable schemas' structure [4]. Although database reverse engineering has been investigated recently, it was not researched in the context of multi-database mining. By taking a unified view of multi-database mining and database reverse engineering, Zhong et al. proposed the RVER (Reverse Variant Entity-Relationship) model to represent the result of multi-database mining [26]. The RVER model can be regarded as a variant of semantic networks that are a kind of well-known method for knowledge representation. From this point of view, multi-database mining can be regarded as a kind of database reverse engineering.

3 Granular Computing (GrC)

The concept of information granulation was first introduced by Zadeh [22]. Information granulation is very essential to human problem solving, and hence has a very significant impact on the design and implementation of intelligent systems. Zadeh [23] identified three basic concepts that underlie human cognition, namely, granulation, organization, and causation. Granulation involves decomposition of whole into parts, organization involves integration of parts into whole, and causation involves association of causes and effects.

An underlying idea of Granular Computing (GrC) uses of groups, classes, or clusters of elements called granules [22,23]. In some situations, although detailed information may be available, it may be sufficient to use granules in order to have an efficient and practical solution. Very precise solutions may in fact not be required for many practical problems. The basic ideas of crisp information granulation have appeared in related fields, such as interval analysis, quantization, rough set theory, Dempster-Shafer theory of belief functions, divide and conquer, cluster analysis, machine learning, databases, and many others. Roughly speaking, GrC is a superset of the theory of fuzzy information granulation, rough set theory and interval computations, and is a subset of granular mathematics [23]. In other words, GrC may be considered as a label of theories, methodologies, techniques, and tools that make use of granules in the process of problem solving.

GrC has many potential applications in knowledge discovery and data mining [20]. There are at least three fundamental issues in GrC: granulation of the universe, description of granules, and relationships between granules. These issues have been considered either explicitly or implicitly in many fields, such as data and cluster analysis, concept formation, and knowledge discovery and data mining. Granulation of a universe involves the decomposition of the universe into parts, or the grouping of individual elements into classes, based on available information and knowledge. Elements in each granule may be interpreted as instances of a concept. They are drawn together by indistinguishability, similarity, proximity or functionality [22,23]. An important function of knowledge discovery and data mining is to establish relationships between granules, such as association and causality.

A challenge in multi-database mining is heterogeneity among multiple databases. That is, no explicit foreign key relationships exist among them usually since different databases may use different terminology and conceptual level to define their scheme. Hence, the key issue is how to find/create the relevance among different databases. Granular computing techniques provide a useful tool to find/create the relevance and association among different databases by changing information granularity [26,8,23,20].

4 Interestingness and Peculiarity

Generally speaking, hypotheses (knowledge) generated from databases can be divided into the following three types: incorrect hypotheses, useless hypotheses, and *new, surprising, interesting hypotheses.* The purpose of data mining is to discover new, surprising, interesting knowledge hidden in databases. Hence, the evaluation of interestingness (including peculiarity, surprisingness, unexpectedness, usefulness, novelty) should be done in pre-processing and/or post-processing of the knowledge discovery process [5,6,11,26]. Here, "evaluating in pre-processing" is to select interesting data before hypotheses generation; "evaluating in post-processing" is to select interesting rules after hypotheses generation. Furthermore, interestingness evaluation may be either *subjective* or *objective* [12]. Here, "subjective" means user-driven, that is, asking the user to explicitly specify what type of data (or rules) are interesting and uninteresting, and the system then generates or retrieves those matching rules; "objective" means data-driven, that is, analyzing structure of data (or rules), predictive performance, statistical significance, and so forth.

Zhong et al. proposed *peculiarity rules* as a new class of rules [26]. A peculiarity rule is discovered from peculiar data by searching the relevance among the peculiar data. Roughly speaking, data are *peculiar* if they represent a peculiar case described by a relatively low number of objects and are very different from other objects in a dataset. Although it looks like the exception rule from the viewpoint of describing a relatively low number of objects, the semantic of the peculiarity rule is with common-sense, which is a feature of the ordinary association rule [1,15].

Illustrative Example. The following rule is a peculiarity one that can be discovered from a relation called *Supermarket-Sales* (see Table 1) in a *Supermarket-Sales* database:

Table 1. Supermarket-Sales

Addr.	Date	meat-sale	vegetable-sale	fruits-sale	...	turnover
Ube	July-1	400	300	450	...	2000
...	July-2	420	290	460	...	2200
...	...	...	...	...	...	...
...	**July-30**	**12**	**10**	**15**	...	**100**
...	July-31	430	320	470	...	2500
...	...	...	...	...	...	...

$rule_1$: *meat-sale(low)* $\wedge$ *vegetable-sale(low)* $\wedge$

fruits-sale(low) $\rightarrow$ *turnover(very-low).*

We can see that this rule just covers data in one tuple on July-30 and its semantic is with common-sense. Hence, algorithms for mining association rules and exception rules may fail to find such useful rules. However, a manager of the supermarket may be interested in such rule because it shows that the turnover was a marked drop.

In order to discover such peculiarity rule, we first need to search peculiar data in the relation *Supermarket-Sales.* From Table 1, we can see that the values of the attributes *meat-sale*, *vegetable-sale*, and *fruits-sale* on July-30 are very different from other values in the attributes. Hence, the values are regarded as peculiar data. Furthermore, $rule_1$ is generated by searching the relevance among the peculiar data. Note that we use the qualitative representation for the quantitative values in the above rules. The transformation of quantitative to qualitative values can be done by using the following background knowledge on information granularity:

Basic granules:
bg_1 = {*high, low, very-low*};
bg_2 = {*large, small, very-small*};
bg_3 = {*many, little, very-little*};
....

Specific granules:
kanto-area = {*Tokyo, Tiba, Saitama, ...*};
chugoku-area = {*Yamaguchi, Hiroshima, Shimane, ...*};
yamaguchi-prefecture = {*Ube, Shimonoseki, ...*};
....

That is, *meat-sale* = *12*, *vegetable-sale* = *10*, *fruits-sale* = *15* and *turnover* = *100* on July 30 are replaced by the granules, "low" and "very-low", respectively.

5 Peculiarity Oriented Mining

The main task of mining peculiarity rules is the identification of peculiarity data. According to our previous papers [26,28], peculiarity data are a subset of objects in the database and are characterized by two features: (1) very different from other objects in a dataset, and (2) consisting of a relatively low number of objects.

There are many ways of finding the peculiar data. In this section, we describe an attribute-oriented method.

5.1 Finding the Peculiar Data

Table 2 shows a relation with attributes A_1, A_2, ..., A_m. In Table 2, let x_{ij} be the ith value of A_j, and n the number of tuples. The peculiarity of x_{ij} can be evaluated by the *Peculiarity Factor*, $PF(x_{ij})$,

$$PF(x_{ij}) = \sum_{k=1}^{n} \sqrt{N(x_{ij}, x_{kj})}. \qquad (1)$$

It evaluates whether x_{ij} occurs in relatively low number and is very different from other data x_{kj} by calculating the sum of the square root of the conceptual distance (N) between x_{ij} and x_{kj}. The reason why the square root is used in Eq. (1) is that we prefer to evaluate closer distances for a relatively large number of data so that the peculiar data can be found from a relatively low number of data.

Table 2. A sample table (relation)

A_1	A_2	...	A_j	...	A_m
x_{11}	x_{12}	...	x_{1j}	...	x_{1m}
x_{21}	x_{22}	...	x_{2j}	...	x_{2m}
⋮	⋮		⋮		⋮
x_{i1}	x_{i2}	...	x_{ij}	...	x_{im}
⋮	⋮		⋮		⋮
x_{n1}	x_{n2}	...	x_{nj}	...	x_{nm}

Major merits of the method are the following:

- It can handle both the continuous and symbolic attributes based on a unified semantic interpretation;
- Background knowledge represented by binary neighborhoods can be used to evaluate the peculiarity if such background knowledge is provided by a user.

If X is a continuous attribute and no background knowledge is available, in Eq. (1),

$$N(x_{ij}, x_{kj}) = |x_{ij} - x_{kj}|. \qquad (2)$$

Table 3 shows the calculation of peculiarity factor. If X is a symbolic attribute and the background knowledge for representing the conceptual distances between x_{ij} and x_{kj} is provided by a user, the peculiarity factor is calculated by the conceptual distances, $N(x_{ij}, x_{kj})$ [8,20,26,28]. However, the conceptual distances are assigned to 1 if no background knowledge is available.

There are two major methods for testing if the peculiar data exist or not (it is called *selection of peculiar data*) after the evaluation for the peculiarity factors. The first is based on a threshold value as shown in Eq. (3),

$$\begin{aligned} threshold = {} & mean\ of\ PF(x_{ij}) + \alpha \times \\ & standard\ deviation\ of\ PF(x_{ij}), \end{aligned} \qquad (3)$$

Table 3. An example of peculiarity factors for a continuous attribute

Region	ArableLand		PF
Hokkaido	**1209**		*134.1*
Tokyo	12		60.9
Osaka	18	$\Longrightarrow$	60.3
Yamaguchi	162		60.5
Okinawa	147		59.4

where α can be adjusted by a user, and $\alpha = 1$ as default. The threshold indicates that a data is a peculiar one if its PF value is much larger than the mean of the PF set. In other words, if $PF(x_{ij})$ is over the threshold value, x_{ij} is a peculiar data.

One can observe that peculiar data can be selected objectively by means of the threshold we are proposing, and the subjective factor (preference) of a user can also be included in the threshold value by adjusting the α.

The other method for *selection of peculiar data* uses the chi-square test that is useful when the data size is sufficiently large [3].

5.2 Attribute Oriented Clustering

Searching the data for a structure of natural clusters is an important exploratory technique. Clusters can provide an informal means of assessing interesting and meaningful groups of peculiar data.

Attribute oriented clustering is used to quantize continuous values, and eventually perform conceptual abstraction [24]. In the real world, there are many real-valued attributes as well as symbolic-valued attributes. In order to discover the better knowledge, conceptual abstraction and generalization are also necessary. Therefore, *attribute oriented clustering* is a useful technique as a step of the peculiarity oriented mining process.

It is a key issue how to do clustering in the environment in which background knowledge on information granularity can either be used or not according to whether such background knowledge exists. Our approach is to provide various methods in the mining process so that the different data can be handled effectively.

If background knowledge on information granularity as stated in Section 4 is available, it is used for conceptual abstraction (generalization) and/or clustering.

If no such background knowledge is available, the *nearest neighbor method* is used for clustering of continuous-values attributes [7].

5.3 An Algorithm

Based on the above-stated preparation, an algorithm of finding peculiar data can be outlined as follows:

Step 1. Execute attribute oriented clustering for each attribute, respectively.
Step 2. For attributes 1 to n do
- *Step 2.1.* Calculate the peculiarity factor $PF(x_{ij})$ in Eq. (1) for all values in an attribute.
- *Step 2.2.* Calculate the threshold value in Eq. (3) based on the peculiarity factor obtained in *Step* 2.1.
- *Step 2.3.* Select the data that are over the threshold value as the peculiar data.
- *Step 2.4.* If the current peculiarity level is enough, then goto *Step* 3.
- *Step 2.5.* Remove the peculiar data from the attribute and thus, we get a new dataset. Then go back to *Step* 2.1.

Step 3. Change the granularity of the peculiar data by using background knowledge on information granularity if the background knowledge is available.

Furthermore, the algorithm can be done in a parallel-distributed mode for multiple attributes, relations and databases because this is an attribute-oriented finding method.

5.4 Relevance Among the Peculiar Data

A peculiarity rule is discovered from the peculiar data, which belong to a cluster, by searching the relevance among the peculiar data. Let $X(x)$ and $Y(y)$ be the peculiar data found in two attributes X and Y respectively. We deal with the following two cases:

- If both $X(x)$ and $Y(y)$ are symbolic data, the relevance between $X(x)$ and $Y(y)$ is evaluated by

$$R_1 = P_1(X(x)|Y(y))P_2(Y(y)|X(x)), \tag{4}$$

  that is, the larger the product of the probabilities of P_1 and P_2 is, the stronger the relevance between $X(x)$ and $Y(y)$ is.
- If both $X(x)$ and $Y(y)$ are continuous attributes, the relevance between $X(x)$ and $Y(y)$ is evaluated by using the method developed in our KOSI system that is a functional relationship finding one [25].

Furthermore, Eq. (4) is suitable for handling more than two peculiar data found in more than two attributes if $X(x)$ (or $Y(y)$) is a granule of the peculiar data.

6 Peculiarity Oriented Multi-Database Mining

Building on the preparatory in the previous sections, this section extends peculiarity oriented mining into multi-database mining.

6.1 Peculiarity Oriented Mining in Multiple Databases

Generally speaking, the tasks of multi-database mining for the first two levels stated in Section 1 can be described as follows:

First, the concept of a foreign key in the relational databases needs to be extended into a foreign link because we are also interested in getting to non-key attributes for data mining from multiple relations in a database [18]. A major work is to find peculiar data in multiple relations for a given discovery task while foreign link relationships exist. In other words, our task is to select n relations, which contain the peculiar data, among m relations ($m \geq n$) with foreign links.

The method for selecting n relations among m relations can be divided into the following steps:

Step 1. Focus on a relation as the *main table* and find the peculiar data from this table. Then elicit the peculiarity rules from the peculiar data by using the methods stated in Section 5.

Step 2. Find the value(s) of the focused key corresponding to the mined peculiarity rule (or peculiar data) in *Step* 1 and change its granularity of the value(s) of the focused key if the background knowledge on information granularity is available.

Step 3. Find the peculiar data in the other relations (or databases) corresponding to the value (or its granule) of the focused key.

Step 4. Select n relations that contain the peculiar data, among m relations ($m \geq n$). In other words, we just select the relations that contain peculiar data relevant to the peculiarity rule mined from the main table.

A peculiarity rule can be discovered from peculiar data hidden in multiple relations by searching the relevance among the peculiar data. If the peculiar data, $X(x)$ and $Y(y)$, are found in two different relations, we need to use a value (or its granule) in a key (or foreign key/link) as the relevance factor, $K(k)$, to find the relevance between $X(x)$ and $Y(y)$. Thus, the relevance between $X(x)$ and $Y(y)$ is evaluated by

$$R_2 = P_1(K(k)|X(x))P_2(K(k)|Y(y)). \tag{5}$$

Furthermore, the above-stated methodology can be extended for mining from multiple databases. A challenge in multi-database mining is a semantic heterogeneity among multiple databases because usually no explicit foreign key/link relationships exists among them. Hence, the key issue of the extension is how to find/create the relevance among different databases. In our methodology, we use *granular computing* techniques based on semantics, approximation, and abstraction for solving the issue [8,23].

We again use the illustrative example mentioned at Section 4. If a manager of the supermarket found that the turnover was a marked drop in one day from a supermarket-sale database, he/she may not understand the deeper reason. Although $rule_1$ as a peculiarity rule (see Section 4) can be discovered from the supermarket-sales database, the deeper reason why the turnover was a marked drop is not explained well. However, if we search several related data sources such as a weather database as shown in Table 4, we can find that there was a violent typhoon that day. Hence, we can understand the deeper reason why the turnover was a marked drop. For this case, the granule of *addr.* = *Ube* in Table 1 needs to be changed into *region* = *Yamaguchi* for creating explicit foreign link between the supermarket-sales database and the weather database. This example will be further described in the next section.

Table 4. Weather

Region	Date	...	Weather
Yamaguchi	July-1	...	sunny
...	July-2	...	cloud
...	...	...	...
...	**July-30**	...	**typhoon (no. 2)**
...	July-31	...	cloud
...	...	...	...

6.2 Representation and Re-learning

We use the RVER (Reverse Variant Entity-Relationship) model to represent the peculiar data and the conceptual relationships among the peculiar data discovered from multiple relations (databases) [26]. Figure 1 shows the general framework of the RVER model. In this figure, the "main entity" is the main table/database specified by a user and the "selected relation" is the table/database with the peculiar data corresponding to the mined peculiarity rule (or peculiar data) in the main table/database. The RVER model can be regarded as a variant of semantic networks that are a kind of well-known method for knowledge representation. From this point of view, multi-database mining can be regarded as a kind of database reverse engineering.

Figure 2 shows the results mined from two databases on supermarket sales at Yamaguchi prefecture and the weather of Japan. The point of which the RVER model is different from an ordinary ER model is that we just represent the attributes that are relevant to the peculiar data and the related peculiar data (or their granules) in the RVER model. Thus, the RVER model provides all interesting information that is relevant to some focusing (e.g. *turnover* = *very-low*, *region* = *Yamaguchi*, and *date* = *July-30* in the supermarket-sale database) for learning more interesting rules among multiple relations (databases).

Re-learning means learning more interesting rules from the RVER model. For example, the following rule can be learned from the RVER model shown in Figure 2:

$rule_2$: *weather(typhoon)* $\rightarrow$ *turnover(very-low)*.

We can see that a manager of the supermarket may be more interested in $rule_2$ (rather than $rule_1$) because $rule_2$ shows a deeper reason why the turnover was a marked drop.

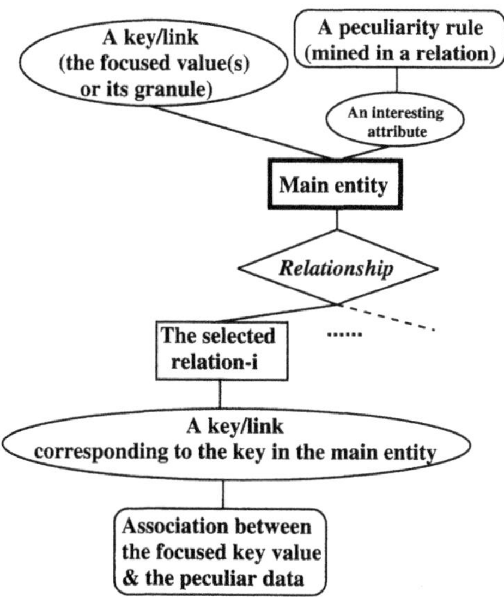

Fig. 1. The frame of the RVER model

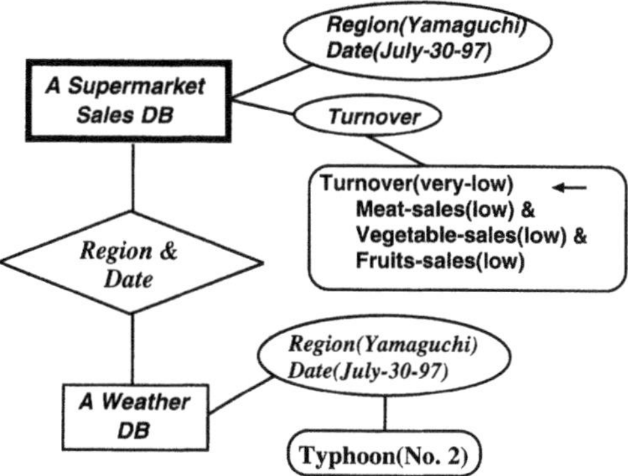

Fig. 2. The RVER model mined from two databases

7 Application in Amino-acid Data Mining

This section discusses a result of mining from the amino-acid data set [17]. The amino-acid data set can be divided into two data sources: amino-acid matrix (including VH and VL amino-acid matrixes) and experimental data (including combining coefficients and coefficients related thermodynamics). The main features of the data set can be summarized as follows:

- The number of the attributes is quite many. That is, the number of the attributes with respect to the amino-acid matrix is 230, the number of the attributes with respect to experimental data is 7.
- The number of the instances is relatively small and only small number of data in the amino-acid matrix changes.

The objective of data mining is to find the association between the amino-acid matrix and experimental data. That is, how experimental data change when amino-acid data are changed.

At first, we find the peculiar data in all attributes respectively by using the method stated in Section 5. As a result, the data denoted in a bold type style in Table 5 (tuples 1-22), Table 6 (tuples 24-35), and Table 7 (tuples 23 and 26) are peculiar data. Note that in the tables, the tuples with the same ID in every table correspond to the same object, and the last tuple T corresponds to the threshold calculated in Eq. (3) with $\alpha = 1$.

From the tuple 23 (i.e. No 23) in Tables 5, 6 and 7, we can see that the value 42 in the attribute Ka (combining coefficients) is a peculiar data and the maximum one in Ka (Table 7), and no any change in the amino-acid matrix. Therefore, we focus on the attribute Ka and search the minimum value in Ka. In other words, we want to find *how coefficients related thermodynamics and the amino-acid matrix change when combining coefficients have big change.* We found that

- The value 0.04 in the attribute Ka (the tuple 26, i.e. No 26) is the minimum one;
- In the same tuple (the tuple 26, i.e. No 26), the values related thermodynamics: -32.6 in *DG*, -53.4 in *DH*, -0.92 in *DCp* are peculiar data;
- The value a in 32 of VL amino-acid matrix is also a peculiar one.

Furthermore, we found that there is a functional relationship between Ka and DG [25]. Therefore, we just use one attribute, Ka or DG, when generating a peculiarity rule.

Figure 3 shows the RVER model mined from two data sources on amino-acid matrix and experimental data. We can see that the RVER model provides all interesting information for further learning more interesting rules among multiple data sources.

Furthermore, the following rules can be learned from the RVER model.

If the value in 32 of VL amino-acid matrix is changed to **a**,
Then the value of Ka *is the minimum one and the values of* DH *and* DCp *are peculiar ones.*

or

If the value of Ka *is the minimum one and the values of* DH *and* DCp *are peculiar ones,*
Then the value in 32 of VL amino-acid matrix is changed to **a**.

The result has been evaluated by an expert [17]. According to his opinion, the discovered rules are reasonable and interesting.

We argue that the peculiarity rules represent a typically unexpected, interesting regularity hidden in the amino-acid data set. The rules are peculiar ones rather than exceptions because of semantic common sense.

Table 5. VH amino-acid matrix with peculiar data and their PF values

No	31	PF(x)	32	PF(x)	33	PF(x)	50	PF(x)	53	PF(x)	56	PF(x)	58	PF(x)	98	PF(x)	99	PF(x)
1	**a**	**34**	d	4	y	4	y	3	y	4	s	1	y	3	w	1	d	2
2	s	1	**a**	**33**	y	4	y	3	y	4	s	1	y	3	w	1	d	2
3	s	1	**e**	**34**	y	4	y	3	y	4	s	1	y	3	w	1	d	2
4	s	1	**n**	**34**	y	4	y	3	y	4	s	1	y	3	w	1	d	2
5	s	1	d	4	**a**	**34**	y	3	y	4	s	1	y	3	w	1	d	2
6	s	1	d	4	**l**	**34**	y	3	y	4	s	1	y	3	w	1	d	2
7	s	1	d	4	**f**	**34**	y	3	y	4	s	1	y	3	w	1	d	2
8	s	1	d	4	**w**	**34**	y	3	y	4	s	1	y	3	w	1	d	2
9	s	1	d	4	y	4	**a**	**34**	y	4	s	1	y	3	w	1	d	2
10	s	1	d	4	y	4	**l**	**34**	y	4	s	1	y	3	w	1	d	2
11	s	1	d	4	y	4	**f**	**34**	y	4	s	1	y	3	w	1	d	2
12	s	1	d	4	y	4	y	3	**a**	**34**	s	1	y	3	w	1	d	2
13	s	1	d	4	y	4	y	3	**l**	**34**	s	1	y	3	w	1	d	2
14	s	1	d	4	y	4	y	3	**p**	**34**	s	1	y	3	w	1	d	2
15	s	1	d	4	y	4	y	3	**w**	**34**	s	1	y	3	w	1	d	2
16	s	1	d	4	y	4	y	3	y	4	**a**	**34**	y	3	w	1	d	2
17	s	1	d	4	y	4	y	3	y	4	s	1	**a**	**34**	w	1	d	2
18	s	1	d	4	y	4	y	3	y	4	s	1	**l**	**34**	w	1	d	2
19	s	1	d	4	y	4	y	3	y	4	s	1	**f**	**34**	w	1	d	2
20	s	1	d	4	y	4	y	3	y	4	s	1	y	3	**a**	**34**	d	2
21	s	1	d	4	y	4	y	3	y	4	s	1	y	3	w	1	**a**	**33**
22	s	1	**a**	**33**	y	4	y	3	y	4	s	1	y	3	w	1	**a**	**33**
23	s	1	d	4	y	4	y	3	y	4	s	1	y	3	w	1	d	2
24	s	1	d	4	y	4	y	3	y	4	s	1	y	3	w	1	d	2
25	s	1	d	4	y	4	y	3	y	4	s	1	y	3	w	1	d	2
26	s	1	d	4	y	4	y	3	y	4	s	1	y	3	w	1	d	2
27	s	1	d	4	y	4	y	3	y	4	s	1	y	3	w	1	d	2
28	s	1	d	4	y	4	y	3	y	4	s	1	y	3	w	1	d	2
29	s	1	d	4	y	4	y	3	y	4	s	1	y	3	w	1	d	2
30	s	1	d	4	y	4	y	3	y	4	s	1	y	3	w	1	d	2
31	s	1	d	4	y	4	y	3	y	4	s	1	y	3	w	1	d	2
32	s	1	d	4	y	4	y	3	y	4	s	1	y	3	w	1	d	2
33	s	1	d	4	y	4	y	3	y	4	s	1	y	3	w	1	d	2
34	s	1	d	4	y	4	y	3	y	4	s	1	y	3	w	1	d	2
35	s	1	d	4	y	4	y	3	y	4	s	1	y	3	w	1	d	2
T		7.44		16.75		16.97		14.34		16.97		7.44		14.34		7.44		10.97

8 Conclusion

A method of mining peculiarity rules from multiple data sources was presented. It is an effective technique of information fusion in data mining to

Table 6. VL amino-acid matrix with peculiar data and their PF values

No	31	PF(x)	32	PF(x)	50	PF(x)	53	PF(x)	91	PF(x)	92	PF(x)	96	PF(x)
1	n	2	n	2	y	2	q	2	s	1	n	2	y	1
2	n	2	n	2	y	2	q	2	s	1	n	2	y	1
3	n	2	n	2	y	2	q	2	s	1	n	2	y	1
4	n	2	n	2	y	2	q	2	s	1	n	2	y	1
5	n	2	n	2	y	2	q	2	s	1	n	2	y	1
6	n	2	n	2	y	2	q	2	s	1	n	2	y	1
7	n	2	n	2	y	2	q	2	s	1	n	2	y	1
8	n	2	n	2	y	2	q	2	s	1	n	2	y	1
9	n	2	n	2	y	2	q	2	s	1	n	2	y	1
10	n	2	n	2	y	2	q	2	s	1	n	2	y	1
11	n	2	n	2	y	2	q	2	s	1	n	2	y	1
12	n	2	n	2	y	2	q	2	s	1	n	2	y	1
13	n	2	n	2	y	2	q	2	s	1	n	2	y	1
14	n	2	n	2	y	2	q	2	s	1	n	2	y	1
15	n	2	n	2	y	2	q	2	s	1	n	2	y	1
16	n	2	n	2	y	2	q	2	s	1	n	2	y	1
17	n	2	n	2	y	2	q	2	s	1	n	2	y	1
18	n	2	n	2	y	2	q	2	s	1	n	2	y	1
19	n	2	n	2	y	2	q	2	s	1	n	2	y	1
20	n	2	n	2	y	2	q	2	s	1	n	2	y	1
21	n	2	n	2	y	2	q	2	s	1	n	2	y	1
22	n	2	n	2	y	2	q	2	s	1	n	2	y	1
23	n	2	n	2	y	2	q	2	s	1	n	2	y	1
24	**a**	**34**	n	2	y	2	q	2	s	1	n	2	y	1
25	**d**	**34**	n	2	y	2	q	2	s	1	n	2	y	1
26	n	2	**a**	**34**	y	2	q	2	s	1	n	2	y	1
27	n	2	**d**	**34**	y	2	q	2	s	1	n	2	y	1
28	n	2	n	2	**a**	**34**	q	2	s	1	n	2	y	1
29	n	2	n	2	**f**	**34**	q	2	s	1	n	2	y	1
30	n	2	n	2	y	2	**a**	**34**	s	1	n	2	y	1
31	n	2	n	2	y	2	**e**	**34**	s	1	n	2	y	1
32	n	2	n	2	y	2	q	2	**a**	**34**	n	2	y	1
33	n	2	n	2	y	2	q	2	s	1	**d**	**34**	y	1
34	n	2	n	2	y	2	q	2	s	1	**a**	**34**	y	1
35	n	2	n	2	y	2	q	2	s	1	n	2	**f**	**34**
T		11.26		11.26		11.26		11.26		7.44		11.26		7.44

Table 7. Experimental data and their PF values

No	Ka x107/M-1	PF(x)	DG/kJmol-1	PF(x)	DH/kJmol-1	PF(x)	TDS/kJmol-1	PF(x)	DCp/kJmol-1K-1	PF(x)
1	9.6	74.38	-46.3	44.2	-97.9	119.79	-51.6	113.6	-2.25	19.67
2	10	74.85	-46.4	43.95	-112.9	145.52	-66.5	142.47	-2.15	19.71
3	16.9	92.03	-47.7	49.28	-108.7	136.72	-61	129.63	-2.26	20.02
4	22	108.06	-48.5	54.07	-115.8	154.33	-67.3	144.9	-2.25	19.67
5	ND	0	ND	0	ND	0	ND	0	ND	0
6	1.5	83.08	-41.4	63.93	-67.3	138.89	-25.9	123.67	-1.8	18.15
7	7.1	75.62	-45.6	46.3	-73.2	131.2	-27.6	121.36	-1.81	18.19
8	2.3	80.84	-42.6	58.04	-65.6	141.15	-23	128.26	-1.78	18.19
9	ND	0	ND	0	ND	0	ND	0	ND	0
10	0.37	86.69	-38	80.07	-53.9	163.18	-15.9	145.42	-1.38	18.4
11	2.9	79.65	-43.1	55.11	-59.8	150.34	-16.7	143.06	-1.27	20.37
12	0.19	87.62	-36.4	87.81	-60.2	149.47	-23.8	126.77	ND	0
13	11	76.25	-46.4	43.95	-81.9	123	-35.4	114.99	-1.43	18.35
14	15	85.34	-47.2	45.93	-84.4	119.01	-37.2	113.34	-0.98	23.6
15	13	80.01	-46.8	44.58	-98.6	120.69	-51.8	113.9	-1.38	18.4
16	12	77.05	-47	45.42	-90.5	118.37	-42.5	112.02	-1.4	18.07
17	2.6	80.08	-43.1	55.11	-72.3	132.28	-29.2	119.57	-1	23.32
18	3.4	79.3	-43.5	54.23	-62.3	146.21	-18.8	137.83	-0.92	24.47
19	23	111.74	-48.5	54.07	-85.3	119.49	-36.8	113.47	-1.58	18.19
20	ND	0	ND	0	ND	0	ND	0	ND	0
21	21	105.05	-48.2	52.46	-113.7	147.16	-65.5	140.19	-1.9	18.69
22	4	79.16	-44.1	52.63	-114.2	148.64	-70.1	154.37	-2.1	19.55
23	**42**	**169.11**	-50.2	67.3	-91.5	117.98	-41.3	112.39	-1.4	18.07
24	34	147.06	-49.5	62.16	-106.3	130.92	-56.8	122.24	-2.42	23.28
25	8.8	74.45	-46.1	44.36	-105.8	129.88	-59.7	126.2	-2.31	21.13
26	**0.04**	88.67	**-32.6**	**105.24**	**-53.4**	**164.42**	-20.8	133.17	**-0.92**	**24.47**
27	0.97	84.53	-40.5	68.11	-53.6	163.73	-13.1	154.83	-1.59	18.17
28	0.64	85.6	-39.4	73.35	-84.4	119.01	-45	111.82	-1.02	23.15
29	9.2	74.25	-46.1	44.36	-76	129.01	-30.1	119.02	-1.64	18.26
30	7.8	75.17	-45.7	45.96	-96.8	119.11	-51.1	113.54	-2.25	19.67
31	6.9	75.93	-45.5	46.87	-105	128.96	-59.5	125.84	-2.39	22.67
32	15.6	87.28	-47.5	48	-101.6	125.09	-54.1	118.04	-2.24	19.81
33	14	82.61	-47.2	45.93	-93.6	117.85	-46.4	111.9	-1.4	18.07
34	ND	0	ND	0	ND	0	ND	0	ND	0
35	12	77.05	-46.8	44.58	-94.6	118.08	-47.8	112.52	-1.97	19.04
T		112.48		71.65		164.16		153.35		24.44

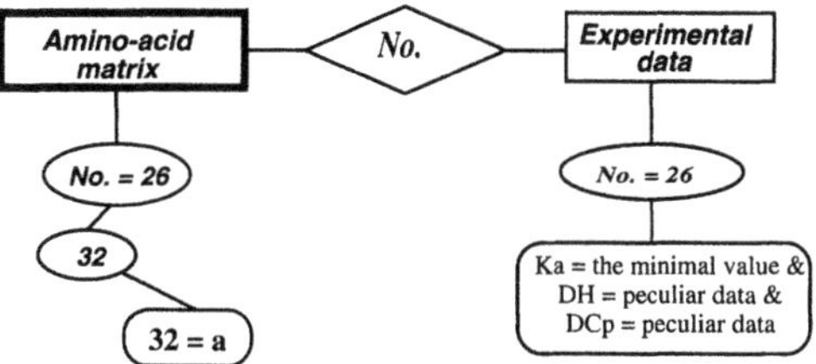

Fig. 3. The RVER model mined from amino-acid data

improve the performance of mining results. This paper showed that such peculiarity rules represent a typically unexpected, interesting regularity hidden in databases.

Here we should mention that Liu's group systematically investigated how to analyze the subjective interestingness of association rules [11,12]. The work of his group is about subjective evaluation of interestingness in post-processing (i.e. evaluating the mined rules). In contrast, our work is about objective evaluation of interestingness in pre-processing (i.e. selecting interesting data (peculiar data) before rule generation). In particular, our approach can mine a new class of patterns: *peculiarity rules* in *multiple* data sources.

So far, as examples with respect to the first two levels of the multi-database mining tasks mentioned in the Introduction, a number of databases such as Japan-survey, amino-acid data, weather, supermarket, web-log, have been tested for our approach. Currently, we are also working on the third level of the multi-database mining task, that is, mining from multiple mixed-media databases [19].

Our future work includes developing a systematic method to mine the rules from multiple data sources where there are no explicitly foreign key (link) relationships, and to induce more interesting rules from the RVER model discovered from multiple data sources by cooperatively using inductive and abductive reasoning.

Acknowledgements

The author would like to thank Prof. Y.Y. Yao, Prof. S. Ohsuga, and Mr. Ohshima for their valuable comments and helps. The author would like to thank Prof. S. Tsumoto and Prof. K. Tsumoto for providing the amino-acid data set and background knowledge, and evaluating the experimental results.

References

1. Agrawal R. et al. (1996) "Fast Discovery of Association Rules", *Advances in Knowledge Discovery and Data Mining* 307-328.

2. Aronis, J.M. et al (1997) "The WoRLD; Knowledge Discovery from Multiple Distributed Databases", *Proc. 10th International Florida AI Research Symposium (FLAIRS-97)* 337-341.
3. Bhattacharyya, G.K. and Johnson, R.A. (1977) *Statistical Concepts and Methods*, John Wiley & Sons.
4. Chiang, Roger H.L. et al (eds.) (1997) "A Framework for the Design and Evaluation of Reverse Engineering Methods for Relational Databases", *Data & Knowledge Engineering*, Vol.21, Elsevier, 57-77.
5. Freitas, A.A. (1998) "On Objective Measures of Rule Surprisingness" J. Zytkow and M. Quafafou (eds.) *Principles of Data Mining and Knowledge Discovery*, LNAI 1510, Springer, 1-9.
6. Hilderman, R.J. and Hamilton, H.J. (2001) "Evaluation of Interestingness Measures for Ranking Discovered Knowledge", D. Cheung, G.J. Williams, Q. Li (Eds) *Advances in Knowledge Discovery and Data Mining*, LNAI 2035, Springer, 247-259.
7. Johnson, R.A. and Wichern, D.W. (1998) *Applied Multivariate Statistical Analysis*, Prentice Hall.
8. Lin, T.Y. (1998) "Granular Computing on Binary Relations 1: Data Mining and Neighborhood Systems", L. Polkowski and A. Skowron (eds.) *Rough Sets in Knowledge Discovery*, Vol. 1, Physica-Verlag, 107-121.
9. Lin, T.Y., Zhong, N., Dong, J., and Ohsuga, S. (1998) "Frameworks for Mining Binary Relations in Data", L. Polkowski and A. Skowron (eds.) *Rough Sets and Current Trends in Computing*, LNAI 1424, Springer, 387-393.
10. Liu, H., Lu H., and Yao, J. (1998) "Identifying Relevant Databases for Multidatabase Mining", X. Wu et al. (eds.) *Research and Development in Knowledge Discovery and Data Mining*, LNAI 1394, Springer, 210-221.
11. Liu, B., Hsu W., and Chen, S. (1997) "Using General Impressions to Analyze Discovered Classification Rules", *Proc. Third International Conference on Knowledge Discovery and Data Mining (KDD-97)*, AAAI Press, 31-36.
12. Liu, B., Hsu W., Chen, S., and Ma, Y. (2000) "Analyzing the Subjective Interestingness of Association Rules", *IEEE Intelligent Systems*, Vol.15, No.5, 47-55.
13. Ribeiro, J.S., Kaufman, K.A., and Kerschberg, L. (1995) "Knowledge Discovery from Multiple Databases", *Proc First Inter. Conf. on Knowledge Discovery and Data Mining (KDD-95)*, AAAI Press, 240-245.
14. Silberschatz, A. and Tuzhilin, A. (1996) "What Makes Patterns Interesting in Knowledge Discovery Systems", *IEEE Trans. Knowl. Data Eng.*, Vol.8, No.6, 970-974.
15. Suzuki E. (1997) "Autonomous Discovery of Reliable Exception Rules", *Proc Third Inter. Conf. on Knowledge Discovery and Data Mining (KDD-97)*, AAAI Press, 259-262.
16. Thrun, S. et al. (Fall 1999) "Automated Learning and Discovery", AI Magazine, 78-82.
17. Tsumoto, K. and Kumagai, I. (2000) "Thermodynamic and Kinetic Analyses of The Antigen-Antibody Interaction Using Mutants", *Research Report of JSAI SIG-KBS-A002*, 83-88.
18. Wrobel, S. (1997) "An Algorithm for Multi-relational Discovery of Subgroups", J. Komorowski et al. (eds.) *Principles of Data Mining and Knowledge Discovery*, LNAI 1263, Springer, 367-375.

19. Wu, J. and Zhong, N. (2001) "An Investigation on Human Multi-Perception Mechanism by Cooperatively Using Psychometrics and Data Mining Techniques", *Proc. 5th World Multi-Conference on Systemics, Cybernetics, and Informatics (SCI-01)*, in Invited Session on Multimedia Information: Managing and Processing, Vol. X, 285-290.
20. Yao, Y.Y. (1999) "Granular Computing using Neighborhood Systems", Roy, R., Furuhashi, T., Chawdhry, P.K. (eds.) *Advances in Soft Computing: Engineering Design and Manufacturing*, Springer, 539-553.
21. Yao, Y.Y. and Zhong, N. (1999) "An Analysis of Quantitative Measures Associated with Rules", N. Zhong and L. Zhou (eds.) *Methodologies for Knowledge Discovery and Data Mining*, LNAI 1574, Springer, 479-488.
22. Zadeh, L.A. (1979) "Fuzzy Sets and Information Granularity", Gupta, N., Ragade, R., and Yager, R. (Eds.) *Advances in Fuzzy Set Theory and Applications*, North-Holland, 3-18.
23. Zadeh, L. A. (1997) "Toward a Theory of Fuzzy Information Granulation and Its Centrality in Human Reasoning and Fuzzy Logic", *Fuzzy Sets and Systems*, Elsevier, 90, 111-127.
24. Zhong, N. and Ohsuga, S. (1994) "Discovering Concept Clusters by Decomposing Databases", *Data & Knowledge Engineering*, Vol.12, No.2, Elsevier, 223-244.
25. Zhong, N. and Ohsuga, S. (1995) "KOSI - An Integrated System for Discovering Functional Relations from Databases", *Journal of Intelligent Information Systems*, Vol.5, No.1, Kluwer, 25-50.
26. Zhong, N., Yao, Y.Y., and Ohsuga, S. (1999) "Peculiarity Oriented Multi-Database Mining", J. Zytkow and J. Rauch (eds.) *Principles of Data Mining and Knowledge Discovery*, LNAI 1704, Springer, 136-146.
27. Zhong, N. (2000) "MULTI-DATABASE MINING: A Granular Computing Approach", *Proc. 5th Joint Conference on Information Sciences (JCIS'00)* in special session on Granular Computing and Data Mining (GrC-DM), 198-201.
28. Zhong, N., Ohshima, M., and Ohsuga, S. (2001) "Peculiarity Oriented Mining and Its Application for Knowledge Discovery in Amino-acid Data", D. Cheung, G.J. Williams, Q. Li (eds.) *Advances in Knowledge Discovery and Data Mining*, LNAI 2035, Springer, 260-269.

Discovery of Temporal Knowledge in Medical Time-Series Databases using Moving Average, Multiscale Matching and Rule Induction

Shusaku Tsumoto[1]

Department of Medicine Informatics, Shimane Medical University, School of Medicine,
89-1 Enya-cho Izumo City, Shimane 693-8501 Japan
email: tsumoto@computer.org

Abstract. Since hospital information systems have been introduced in large hospitals, a large amount of data, including laboratory examinations, have been stored as temporal databases. The characteristics of these temporal databases are: (1) inhomogeneity of each record, (2) a large number of attributes in each record and (3) bias of the number of measurements for patients suffering from severe chrononic diseases. The characteristics of these temporal databases are: Even medical experts cannot deal with these large databases, the interest in mining some useful information from the data are growing. In this paper, we introduce a combination of extended moving average method, multiscale matching and rule induction method to discover new knowledge in medical temporal databases. This method was applied to two medical datasets, the results of which show that interesting knowledge is discovered from each database.

Keywords
temporal data mining, temporal abstraction, rough sets, extended moving average, qualitative trend

1 Introduction

Since hospital information systems have been introduced in large hospitals, a large amount of data, including laboratory examinations, have been stored as temporal databases[18]. For example, in a university hospital, where more than 1000 patients visit from Monday to Friday, a database system stores more than 1 GB numerical data of laboratory examinations. Thus, it is highly expected that data mining methods will find interesting patterns from databases because medical experts cannot deal with those large amount of data.

The characteristics of these temporal databases are: (1) Each record are inhomogeneous with respect to time-series, including short-term effects and long-term effects. (2) Each record has more than 1000 attributes when a patient is followed for more than one year. (3) When a patient is admitted for a long time, a large amount of data is stored in a very short term. Even medical experts cannot deal with these large temporal databases, the interest in mining some useful information from the data is growing.

In this paper, we introduce a rule discovery method, combined with extended moving average method, multiscale matching for qualitative trend to discover new knowledge in medical temporal databases. In this system, extended moving average method and multi-scale matching are used for preprocessing, to deal with irregularity of each temporal data. Using several parameters for time-scaling, given by users, this moving average method generates a new database for each time scale with summarized attributes. For matching time sequences, multiscale matching was applied. Then, rule induction method is applied to each new database with summarized attributes. This method was applied to two medical datasets, the results of which show that interesting knowledge is discovered from each database.

This paper is organized as follows. Section 2 introduces the definition of probabilistic rules. Section 3 discusses the characteristics of temporal databases in hospital information systems. Section 4 presents extended moving average method. Section 5 introduces second preprocessing methods to extract qualitative trend and rule discovery method with qualitative trend. Section 6 shows experimental results. Section 7 gives a brief discusson of the total method. Finally, Section 8 concludes this paper.

2 Probabilistic Rules and Conditional Probabilities

2.1 Definition of Rules based on Rough Sets

Before discussing temporal knowledge discovery, we first discuss the characteristics of probabilistic rules. In this section, we use the following notations introduced by Grzymala-Busse and Skowron[12], which are based on rough set theory[10].

Let U denote a nonempty, finite set called the universe and A denote a nonempty, finite set of attributes, respectively. For an attribute $a \in A$, a can be viewed as a mapping $a : U \to V_a$, where V_a is called the domain of a. Then, a decision table is defined as an information system, $A = (U, A \cup \{d\})$, where d is a decision attribute given by domain experts. The atomic formulae over $B \subseteq A \cup \{d\}$ and V_a are expressions of the form $[a = v]$, called descriptors over B, where $a \in B$ and $v \in V_a$. The set $F(B, V)$ of formulas over B is the least set containing all atomic formulas over B and closed with respect to disjunction, conjunction and negation. For each $f \in F(B, V)$, f_A denote the meaning of f in A, i.e., the set of all objects in U with property f, defined inductively as follows.

1. If f is of the form $[a = v]$ then, $f_A = \{s \in U | a(s) = v\}$
2. $(f \wedge g)_A = f_A \cap g_A$; $(f \vee g)_A = f_A \cup g_A$; $(\neg f)_A = U - f_a$

By the use of the framework above, classification accuracy and coverage, or true positive rate is defined as follows.

Let R and D denote a formula in $F(B, V)$ and a set of objects which belong to a decision d, respectively. Classification accuracy and coverage(true positive rate) for $R \to d$ is defined as:

$$\alpha_R(D) = \frac{|R_A \cap D|}{|R_A|} (= P(D|R)), \text{ and } \kappa_R(D) = \frac{|R_A \cap D|}{|D|} (= P(R|D)),$$

where $|S|$, $\alpha_R(D)(= P(D|R))$, $\kappa_R(D)(= P(R|D))$ and P(S) denote the cardinality of a set S, a classification accuracy of R as to classification of D and coverage (a true positive rate of R to D), and probability of S, respectively. [1]

By the use of accuracy and coverage, a probabilistic rule $R \stackrel{\alpha,\kappa}{\to} d$ is defined as a pair (R,d) such that

1. $R = \wedge_j [a_j = v_k]$,
2. $\alpha_R(D)(= P(D|R)) \geq \delta_\alpha$, and
3. $\kappa_R(D)(= P(R|D)) \geq \delta_\kappa$.

If the thresholds for accuracy and coverage are set to high values, the meaning of the conditional part of probabilistic rules corresponds the highly overlapped region. Figure 1 depicts the Venn diagram of probabilistic rules with highly overlapped region.

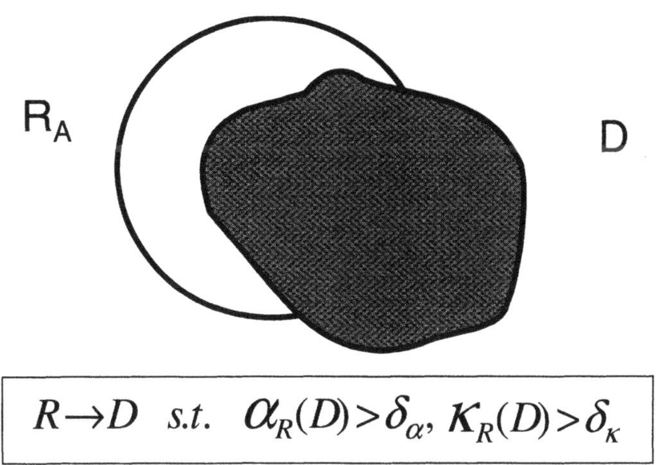

$$R \to D \text{ s.t. } \alpha_R(D) > \delta_\alpha, \kappa_R(D) > \delta_\kappa$$

Fig. 1. Venn Diagram of Probabilistic Rules

[1] These indices are closely related with confidence and support in association rules[3]. Classification accuracy and coverage are equivalent to confidence and support ratio, although the former terminologies have been widely used in the machine learning literature before[7]. In this paper, we adopt the terminology, accuracy and coverage because these names are much more fitted to the meaning of these two indices.

It is notable that $\alpha_R(D)$ measures the degree of the sufficiency of a proposition, $R \rightarrow D$, and that $\kappa_R(D)$ measures the degree of its necessity. For example, if $\alpha_R(D)$ is equal to 1.0, then $R \rightarrow D$ is true. On the other hand, if $\kappa_R(D)$ is equal to 1.0, then $D \rightarrow R$ is true. Thus, if both measures are 1.0, then $R \leftrightarrow D$.

2.2 MDL principle of Accuracy and Coverage

One of the important characteristics of the relation between classification accuracy and coverage is a trade-off relation on description length, as illustrated in Figure 2. When one additional attribute-value pair is included into a given conjunctive formula so that the accuracy of the formula increases, the coverage of the formula will decrease. This relation can be viewed as one variant of MDL principle(Minimum Description Length principle)[11] which is easy to be proved from the definitions of these measures.

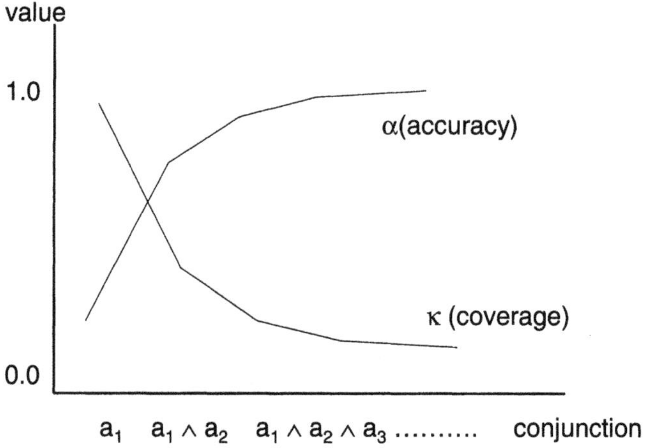

Fig. 2. Trade-off relations between classification accuracy and coverage

Let us define the description length of a rule as:

$$L = -\log_2 \alpha_R(D) - \log_2 \kappa_R(D),$$

which represents the length of a bit strings to describe all the information about classification of accuracy and coverage. In this definition, the length of coverage corresponds to the cost of "theory" in MDL principle because of the following theorem on coverage.

Proposition 1 (Monotonicity of Coverage). *Let $R(j)$ denote an attribute-value pair, which is a conjunction of $R(i)$ and $[a_{i+1} = v_j]$. Then,*

$$\kappa_{R(j)}(D) \leq \kappa_{R(i)}(D).$$

Proof.
Since $R(j)_A \subseteq R(i)_A$ *holds,* $\kappa_{R(j)}(D) = \frac{|R(j)_A \cap D|}{|D|} \leq \frac{|R(i)_A \cap D|}{|D|} = \kappa_{R(i)}(D)$.
□

Then, from their definitions, the following relation will hold unless $\alpha_R(D)$ or $\kappa_R(D)$ is equal to 1.0: [2]

$$\begin{aligned} L &= -\log_2 \alpha_R(D) - \log_2 \kappa_R(D) \\ &= -\log_2 \frac{P(R \cap D)}{P(R)} - \log_2 \frac{P(R \cap D)}{P(D)} \\ &= -\log_2 \frac{(P(R \cap D)P(R \cap D)}{P(D)P(R)} \\ &\geq -\log_2 \frac{P(R)}{P(D)}. \end{aligned}$$

$P(R)$ and $P(D)$ are defined as:

$$P(R) = \frac{|R_A|}{|U|} \quad and \quad P(D) = \frac{|D|}{|U|},$$

where U denotes the total samples.

When we add an attribute-value pair to the conditional part of a rule, the cardinality of $[x]_R$ will decrease and equivalently, the value of $P(R)$ will be smaller. Thus, $\log_2 P(R)$ will approach to $-\infty$ as a result.

Thus, if we want to get a rule of high accuracy, the coverage of this rule will be very small, which causes the high cost of the description of rules. On the other hand, if we want to get a rule of high coverage, the accuracy of this rule will be very small, which also causes the high cost of the description of rules.

It also means that a rule of high accuracy should be described with additional information about positive examples which do not support the rule, or that a rule of high coverage should be described with additional information about negative examples which support the rule.

2.3 Rule Induction Algorithm

Rules are induced from datasets by using the modification of the algorithm introduced in PRIMEROSE[13] and its extention, PRIMEROSE-REX[14], as shown in Figure 3 and Figure 4. The first algorithm shown in Figure 3

[2] Since the MDL principle do not consider the concept of coverage, it is difficult to incorporate the meaning of coverage in an explicit way. However, as discussed in the section on negative rules, the situation when the coverage is equal to 1.0 has a special meaning to express the information about negative reasoning. It will be our future work to study the meaning when the coverage is equal to 1.0. in the context of the description length of "theory".

is a feature selection for probabilistic rules and work as follows. (1)First, it selects a descriptor $[a_i = v_j]$ from the list of attribute-value pairs, denoted by L. (2) Then, it checks whether this descriptor overlaps with a set of positive examples, denoted by D. (3) If so, the algorithm checks whether its coverage is larger than the threshold δ_κ. If the coverage is larger than the threshold, then this descriptor is added to L_ir, the formula for the conditional part of the probabilistic rules of D. (4) Then, $[a_i = v_j]$ is deleted from the list L. This procedure, from (1) to (4) will continue unless L is empty.

Then rules are induced by applying the second algorithm shown in Figure 4.

This algorithm works in the following way. (1) First, it substitutes L_1, which denotes a list of formula composed of only one descriptor, with the list L_{ir} generated by the former algorithm shown in Fig. 3. (2) Then, until L_1 becomes empty, the following procedures will continue: (a) A formula $[a_i = v_j]$ is removed from L_1. (b) Then, the algorithm checks whether $\alpha_R(D)$ is larger than the threshold or not. If so, then this formula is included a list of the conditional part of positive rules. Otherwise, it will be included into M, which is used for a candidates of making conjunctive formulae in the next loop. (3) When L_1 is empty, the next list L_2 is generated from the list M. For further information about these probabilistic rules, reader may refer to [17].

```
procedure Feature Selection;
  var
    L : List;
      /* A list of elementary attribute-value pairs */
  begin
    L := P0;
    /* P0: A list of elementary attribute-value pairs
      given in a database */
    while (L ≠ {}) do
      begin
        Select one pair [a_i = v_j] from L;
        Calculate κ_[a_i=v_j](D);
        if (κ_[a_i=v_j](D) ≥ δ_κ) then
        /* D: positive examples of a target class d */
          L_ir := L_ir + [a_i = v_j];
            /* Candidates for Positive Rules */
    L := L − [a_i = v_j];
      end
  end {Feature Selection};
```

Fig. 3. Feature Selection for Probabilistic Rules

```
procedure Positive Rules;
  var
    i : integer;    M, L_i : List;
    L_ir: A list of candidates generated by
induction of exclusive rules
    S_ir: A list of positive rules
  begin
    L_1 := L_ir;
    /* L_ir: A list of candidates generated by
      induction of exclusive rules */
    i := 1;   M := {};
    for i := 1 to n do
    /* n: Total number of attributes given
      in a database */
      begin
        while ( L_i ≠ {} ) do
          begin
            Select one pair R = ∧[a_i = v_j] from L_i;
            L_i := L_i − {R};
            if  (α_R(D) > δ_α)
              then  do S_ir := S_ir + {R};
          /* Include R in a list of the Positive Rules S_ir */
            else M := M + {R};
          end
        L_{i+1} := (A list of the whole combination
                   of the conjunction formulae in M);
      end
  end {Positive Rules};
```

Fig. 4. Induction of Positive Rules

3 Temporal Databases in Hospital Information Systems

Since incorporating temporal aspects into databases is still an ongoing research issue in database area[1], temporal data are stored as a table in hospital information systems(H.I.S.). Table 1.1 shows a typical example of medical data, which is retrieved from H.I.S. The first column denotes the ID number of each patient, and the second one denotes the date when the datasets in this row is examined. Each row with the same ID number describes the results of laboratory examinations. For example, the second row shows the data of the patient ID 1 on 04/19/1986. This simple database show the following characteristics of medical temporal databases.

(1) The Number of Attributes are too many Even though the dataset of a patient focuses on the transition of each examination (attribute), it would be difficult to see its trend when the patient is followed for a long time. If one wants to see the long-term interaction between attributes, it would be almost impossible. In order to solve this problems, most of H.I.S. systems provide several graphical interfaces to capture temporal trends[18]. However, the interactions among more than three attributes are difficult to be studied even if visualization interfaces are used.

(2) Irregularity of Temporal Intervals Temporal intervals are irregular. Although most of the patients suffering from chronic diseases will come to the hospital every two weeks or one month, physicians may not make laboratory tests at each time. When a patient has a acute fit or suffers from acute diseases, such as pneumonia, laboratory examinations will be made every one to three days. On the other hand, when his/her status is stable, these test may not be made for a long time. Patient ID 1 is a typical example. Between 04/30 and 05/08/1986, he suffered from a pneumonia and was admitted to a hospital. Then, during the therapeutic procedure, laboratory tests were made every a few days. On the other hand, when he was stable, such tests were ordered every one or two years.

(3) Missing Values In addition to irregularity of temporal intervals, datasets have many missing values. Even though medical experts will make laboratory examinations, they may not take the same tests in each instant. Patient ID 1 in Table 1 is a typical example. On 05/06/1986, medical physician selected a specific test to confirm his diagnosis. So, he will not choose other tests. On 01/09/1989, he focused only on GOT, not other tests. In this way, missing values will be observed very often in clinical situations.

Table 1. An Example of Temporal Database

ID	Date	GOT	GPT	LDH	γ-GTP	TP	edema	$\cdots$
1	19860419	24	12	152	63	7.5	-	$\cdots$
1	19860430	25	12	162	76	7.9	+	$\cdots$
1	19860502	22	8	144	68	7.0	+	$\cdots$
1	19860506							$\cdots$
1	19860508	22	13	156	66	7.6	-	$\cdots$
1	19880826	23	17	142	89	7.7	-	$\cdots$
1	19890109	32					-	$\cdots$
1	19910304	20	15	369	139	6.9	+	$\cdots$
2	19810511	20	15	369	139	6.9	-	$\cdots$
2	19810713	22	14	177	49	7.9	-	$\cdots$
2	19880826	23	17	142	89	7.7	-	$\cdots$
2	19890109	32					-	$\cdots$
	$\cdots$							

These characteristics have already been discussed in KDD area[5]. However, in real-world domains, especially domains in which follow-up studies are crucial, such as medical domains, these ill-posed situations will be distinguished. If one wants to describe each patient (record) as one row, then each row have too many attributes, which depends on how many times laboratory examinations are made for each patient. It is notable that although the above discussions are made according to the medical situations, similar situations may occur in other domains with long-term follow-up studies.

3.1 Temporal Knowledge Discovery

Figure 5 illustrates the total process of temporal knowledge discovery in temporal data from hospital information systems. The first preprocessing process applies extended moving average method, which will be discussed in the next section, to temporal databases. Then, the second preprocessing process summarizes temporal sequences as their qualitative trend. Finally, rule induction method is applied to the table with qualitative trend.

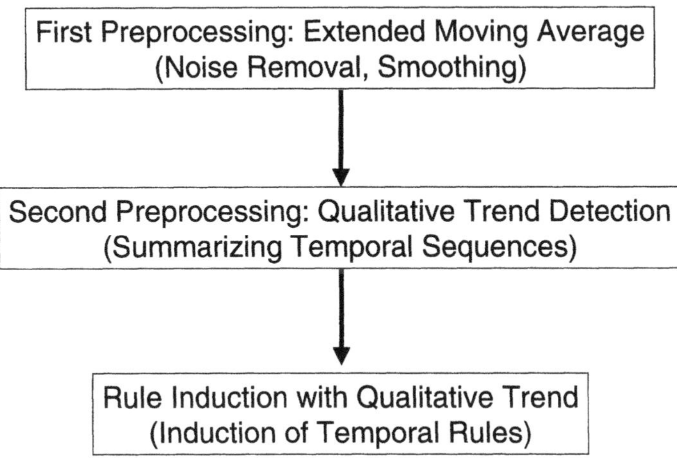

Fig. 5. Temporal Knowledge Discovery

4 First Preprocessing: Extended Moving Average Methods

4.1 Moving Average Methods

Averaging mean methods have been introduced in statistical analysis[6]. Temporal data often suffers from noise, which will be observed as a spike or sharp wave during a very short period, typically at one instant. Averaging mean methods remove such an incidental effect and make temporal sequences smoother.

With one parameter w, called *window*, moving average $\hat{y}_w$ is defined as follows:

$$\hat{y}_w = \sum_{j=1}^{w} y_j.$$

For example, in the case of GOT of patient ID 1, $haty_1$ is calculated as: $\hat{y}_1 = (24 + 25 + 22 + 22 + 22)/5 = 23.0$, where $haty_1$ shows the moveing average of the first window in the above table. It is easy to see that $\hat{y}_w$ will remove the noise effect which continue less than w points.

The advantage of moving average method is that it enables to remove the noise effect when inputs are given periodically[6]. For example, when some tests are measured every several days[3], the moving average method is useful to remove the noise and to extract periodical domains. However, in real-world domains, inputs are not always periodical, as shown in Table 1.1 Thus, when applied time-series are irregular or discrete, ordinary moving average methods are powerless. Another disadvantage of this method is that it cannot be applicable to categorical attributes. In the case of numerical attributes, average can be used as a summarized statistic. On the other hand, such average cannot be defined for categorical attributes.

Thus, we introduce the extended averaging method to solve these two problems in the subsequent subsections.

4.2 Extended Moving Average for Continuous Attributes

In this extension, we first focus on how moving average methods remove noise. The key idea is that a window parameter w is closely related with periodicity. If w is larger, then the periodical behavior whose time-constant is lower than w will be removed. Usually, a spike by noise is observed as a single event and this effect will be removed when w is taken as a large value. Thus, the choice of w separates different kinds of time-constant behavior in each attribute and in the extreme case when w is equal to total number of temporal events, all the temporal behavior will be removed. We refer to this extreme case as $w = \infty$.

The extended moving average method is executed as follows: first calculates y_∞ for an attribute y. Second, the method outputs its maximum and minimum values for the total period.. Then, according to the selected values for w, a set of sequence $\{y_w(i)\}$ for each record is calculated and the minimum and maximum values for each w are output.

For example, if $\{w\ \}$ is equal to {10 years, 5 years, 1 year, 3 months, 2 weeks}, then for each element in $\{w\}$, the method uses the time-stamp attribute for calculation of each $\{y_w(i)\}$ in order to deal with irregularities.

[3] This condition guarantees that measurement is approximately continuous

In the case of Table 1, when w is taken as 1 year, all the rows are aggregated into several components as shown in Table 2 in which each line shows the boundary of the window. For example, the second horizontal line shows the boundary between 1986 and 1987. Since there are no data between 1987 and 1988, the third horizonal line is next to the second one. In this way, we have a window between two horizontal lines in Table 2. Moving average is calulated from each window. In the above example, the moving average value of GOT in the first window is calculated as:

$$\hat{y}_1 = (24 + 25 + 22 + 22)/4 = 23.25.$$

From this aggregation, a sequence y_w for each attribute is calculated as in Table 3. For example, $\hat{y}_\infty$ is caluclated as:

$$\hat{y}_w = (24 + 25 + 22 + 22 + 23 + 32 + 20)/6 = 24.$$

Table 2. Aggregation for w= 1 (year)

ID	Date	GOT	GPT	LDH	γ-GTP	TP	edema	...
1	19860419	24	12	152	63	7.5	-	...
1	19860430	25	12	162	76	7.9	+	...
1	19860502	22	8	144	68	7.0	+	...
1	19860506							...
1	19860508	22	13	156	66	7.6	-	...
1	19880826	23	17	142	89	7.7	-	...
1	19890109	32					-	...
1	19910304	20	15	369	139	6.9	+	...
	...							

The selection of w can be automated. The simpliest way to calculate w is to use the power of natural number, such as 2. For example, we can use 2^n as the window length: 2, 4, 8, 16,..... Using this scale, two weeks, three months, one year correspond to $16 = 2^4$, $64 = 2^6$, $256 = 2^8$.

4.3 Categorical Attributes

One of the disadvantages of moving average method is that it cannot deal with categorical attributes. To solve this problem, we will classify categorical attributes into three types, whose information should be given by users. The first type is *constant*, which will not change during the follow-up period. The second type is *ranking*, which is used to rank the status of a patient. The third type is *variable*, which will change temporally, but ranking is not useful. For

Table 3. Moving Average for w= 1 (year)

ID	Window	GOT	GPT	LDH	γ-GTP	TP	edema	$\cdots$
1	1	23.25	11.25	153.5	68.25	7.5	?	$\cdots$
1	2			N/A				$\cdots$
1	3	23	17	142	89	7.7	?	$\cdots$
1	4	32					?	$\cdots$
1	6			N/A				$\cdots$
1	6			N/A				$\cdots$
1	7	20	15	369	139	6.9	?	$\cdots$
1	∞	24	12.83	187.5	83.5	7.43	?	$\cdots$
	$\cdots$							

the first type, extended moving average method will not be applied. For the second one, integer will be assigned to each rank and extended moving average method for continuous attributes is applied. On the other hand, for the third one, the temporal behavior of attributes is transformed into statistics as follows.

First, the occurence of each category (value) is counted for each window. For example, in Table 1.2, *edema* is a binary attribute and variable. In the first window, an attribute *edema* takes {-,+,+,-}.[4] So, the occurence of $-$ and $+$ are 2 and 2, respectively. Then, each conditional probability will be calculated. In the above example, probabilities are equal to $p(-|w_1) = 2/4$ and $p(+|w_1) = 2/4$. Finally, for each probability, a new attribute is appended to the table (Table 4).

Table 4. Final Table with Moving Average for w= 1 (year)

							edema		
ID	Period	GOT	GPT	LDH	γ-GTP	TP	(+)	(-)	$\cdots$
1	1	23.25	11.25	153.5	68.25	7.5	0.5	0.5	$\cdots$
1	2	23	17	142	89	7.7	0.0	1.0	$\cdots$
1	3	32					0.0	1.0	$\cdots$
1	4						0.0	1.0	$\cdots$
1	5	20	15	369	139	6.9	1.0	0.0	$\cdots$
1	∞	24	12.83	187.5	83.5	7.43	0.43	0.57	$\cdots$
	$\cdots$								

Summary of Extended Moving Average All the process of extended moving average is used to construct a new table for each window parameter as the first preprocessing. Then, second preprocessing method will be applied

[4] Missing values are ignored for counting.

to newly generated tables. The first preprocessing method is summarized as shown in Figure 6.

1. Repeat for each w in List L_w,
   (a) Select an attribute in a List L_a;
      i. If an attribute is numerical, then calculate moving average for w;
      ii. If an attribute is constant, then break;
      iii. If an attribute is rank, then assign integer to each ranking; calculate moving average for w;
      iv. If an attribute is variable, calculate accuracy and coverage of each category;
   (b) If L_a is not empty, goto (a).
   (c) Construct a new table with each moving average.
2. Construct a table for $w = \infty$.

Fig. 6. First Preprocessing Method

5 Second Preprocessing and Rule Discovery

5.1 Summarizing Temporal Sequences

From the data table after processing extended moving average methods, several preprocessing methods may be applied in order for users to detect the temporal trends in each attribute. One way is discretization of time-series by clustering introduced by Das[4]. This method transforms time-series into symbols representing qualitative trends by using a similarity measure. Then, time-series data is represented as a symbolic sequence. After this preprocessing, rule discovery method is applied to this sequential data. Another way is to find auto-regression equations from the sequence of averaging means. Then, these quantitative equations can be directly used to extract knowledge or their qualitative interpretation may be used and rule discovery[3], other machine learning methods[7], or rough set method[13] can be applied to extract qualitative knowledge.

In this research, we adopt two approaches and transforms databases into two forms: one mode is applying temporal abstraction method[9] with multiscale matching[8] as second preprocessing and transforms all continuous attributes into temporal sequences. The other mode is applying rule discovery to the data after the first preprocessing without second one. The reason why we adopted these two mode is that we focus not only on temporal behavior of each attribute, but also on association among several attributes. Although Miksch's method[9] and Das's approach[4] are very efficient to extract knowledge about transition, they cannot focus on association between attributes in an efficient way. For the latter purpose, much simpler rule discovery algorithm are preferred.

5.2 Continuous Attributes and Qualitative Trend

To characterize the deviation and temporal change of continuous attributes, we introduce standardization of continuous attributes. For this, we only needs the total average $\hat{y}_{\infty}$ and its standardization σ_{∞}. With these parameters, standardized value is obtained as:

$$z_w = \frac{y_w - \hat{y}_{\infty}}{\sigma_{\infty}}.$$

The reason why standardization is introduced is that it makes comparison between continuous attributes much easier and clearer, especially, statistic theory guarantees that the coefficients of a auto-regression equation can be compared with those of another equation[6].

After calculating the standardized values, an extraction algorithm for qualitative trends is applied[9] with multiscale matching briefly shown in the next subsection.

This method is processed as follows: First, this method uses data smoothing with window parameters. Secondly, smoothed values for each attributes are classified into seven categories given as domain knowledge about laboratory test values: extremely low, substantially low, slightly low, normal range, slightly high, substantially high, and extremely high. With these categories, qualitative trends are calculated and classified into the following ten categories by using guideline rules: dangerous decrease(D), decrease too fast(A1), normal decrease(A2), decrease too slow(A3), zero change(Z), increase too fast(B1), normal increase(B2), increase too slow(B3), dangerous increase(C). Figure 5.2 and 5.2 illustrates the qualitative trends from D to Z and those from C to Z, respectively. For matching temporal sequences with guideline rules, multiscale matching method is applied. For example, if the value of some laboratory tests change from substantially high to normal range within a very short time, the qualitative trend will be classified into A1(decrease too fast). For further information, please refer to [9].

5.3 Multiscale Matching

Multiscale matching is based on two basic ideas: the first one is to use the curvature of the curve to detect the points of inflection. The second idea is to use the scale factor to calculate the curvature of the smoothed curve[8]. The curvature is given as:

$$c(t) = \frac{y''}{(1 + (y')^2)^{3/2}},$$

where $y' = dy/dt$ and $y'' = d^2y/dt^2$. In order to compute the curvature of the curve at varying levels of detail, the convolution is applied to the function y

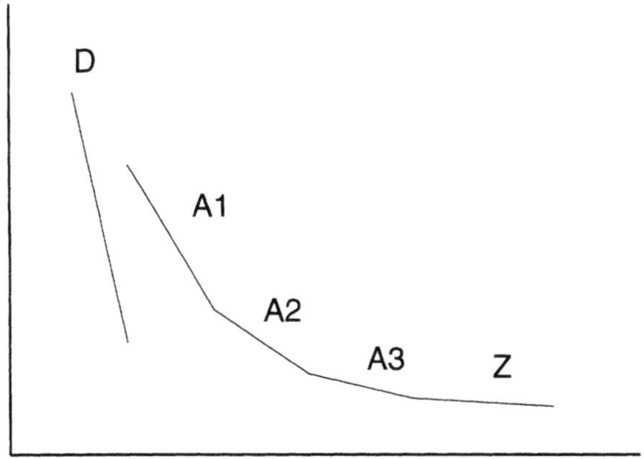

Fig. 7. Qualitative Trend: D to Z

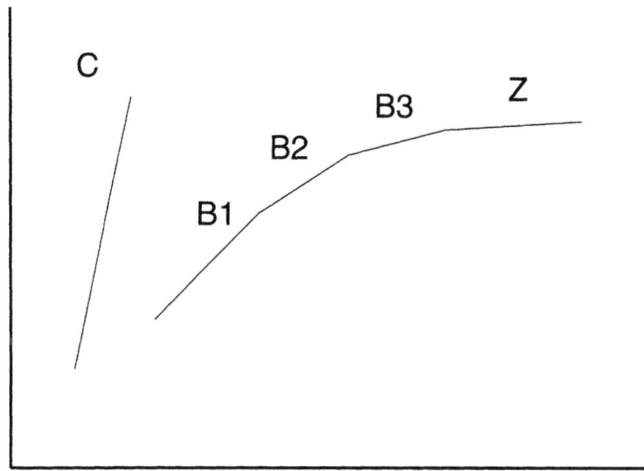

Fig. 8. Qualitative Trend: C to Z

with a one-dimensional Gaussian kernel $g(t, \sigma)$ of the width (scaling factor) σ:

$$g(t, \sigma) = \frac{1}{\sigma\sqrt{2\pi}} e^{-t^2/2\sigma^2}.$$

$Y(t, \sigma)$, the convolution of $y(t)$ is defined as:

$$Y(t, \sigma) = y(t) \otimes g(t, \sigma) = \int_{-\infty}^{\infty} y(t) \frac{1}{\sigma\sqrt{2\pi}} e^{-t^2/2\sigma^2} du.$$

According to the characteristics of the convolution, the derivative and the second derivative is calculated as:

$$Y'(t,\sigma) = y(t) \otimes \frac{\partial g(t,\sigma)}{\partial t} \quad \textit{and} \quad Y''(t,\sigma) = y(t) \otimes \frac{\partial^2 g(t,\sigma)}{\partial t^2}$$

Using $Y(t,\sigma)$, $Y'(t,\sigma)$ and $Y''(t,\sigma)$, we can calculate the curvature of a given curve for each value of σ within one window w:

$$c(t) = \frac{y(t) \otimes \dfrac{\partial^2 g(t,\sigma)}{\partial t^2}}{\left(1 + \left(y(t) \otimes \dfrac{\partial g(t,\sigma)}{\partial t}\right)^2\right)^{3/2}}$$

This gives a sequence of the value of curvature for each time series. If two time series sequence is similar with respect to temporal change, two sequences of curvature will be similar. Furthermore, since we calculate the curvature for each scaling factor, we can compare between these sequences from the local level to global level. For further information, please refer to [8] and [9].

5.4 Rule Discovery Algorithm

For rule discovery, a simple rule induction algorithm discussed in [16] is applied, where continuous attributes are transformed into categorical attributes with a cut-off point. As discussed in Section 3, moving average method will remove the temporal effect shorter than a window parameter. Thus, $w = \infty$ will remove all the temporal effect, so this moving average can be viewed as data without any temporal characteristics. If rule discovery is applied to this data, it will generate rules which represents non-temporal association between attributes. In this way, data after processing w-moving average is used to discover association with w or longer time-effect. Ideally, from $w = \infty$ down to $w = 1$, we decompose all the independent time-effect associations between attributes. However, the time-constant in which users are interested will be limited and the moving average method shown in Section 3 uses a set of w given by users. Thus, application of rule discovery to each table will generate a sequence of temporal associations between attributes. If some temporal associations will be different from associations with $w = \infty$, then these specific relations will be related with a new discovery.

5.5 Summary of Second Preprocessing and Rule Discovery

Second preprocessing method and rule discovery are summarized as shown in Figure 9.

1. Calculate $\hat{y}_\infty$ and σ_∞ from the table of $w = \infty$;
2. Repeat for each w in List L_w;
   (w is sorted in a descending order.)
   (a) Select a table of w: T_w;
      i. Standardize continuous and ranking attributes;
      ii. Calculate qualitative trends for continuous and ranking attributes with multiscale matching;
      iii. Construct a new table for qualitative trends;
      iv. Apply rule discovery method for temporal sequences;
   (b) Apply rule induction methods to the original table T_w;

Fig. 9. Second Preprocessing and Rule Discovery

6 Experimental Results

The above rule discovery system is implemented in CEARI(Combination of Extended Moving Average and RUle Induction). CEARI was applied to two clinical databases. One is on motor neuron diseases, which consists of 1477 samples, 3 classes. Each patient is followed during 15 years. The other one is on cerebrovascular diseases (CVD), which has 2610 records, described by 12 classes. Each record followed up at least 10 years and the averaged number of attributes are 2715.

A list of w, $\{w\}$ was set to $\{$10 years, 5 years, 1 year, 3 months, 2 weeks$\}$ and thresholds, δ_α and δ_κ were set to 0.60 and 0.30,respectively.

6.1 Discovered Results in MND

One of the most interesting problems of Motor neuron diseases (MND) is how long it takes each patient to suffer from respiratory failure, which is the main cause of death.[5] It is empirically known that some types of MND is more progressive than other types and that their survival period is much shorter than others. The database for this analysis describes all the data of patients suffering from MND.

Non-temporal Knowledge The most interesting discovered rules are:

$$[Major\_Pectoralis < 3] \rightarrow [PaCO_2 > 50] \; (Accuracy: 0.87, Coverage: 0.57),$$
$$[Minor\_Pectoralis < 3] \rightarrow [PaO_2 < 61] \quad (Accuracy: 0.877, Coverage: 0.65).$$

Both rules mean that if some of the muscles of chest, called Major Pectoralis and Minor Pectoralis are weak, then respiratory function is low, which suggests that muscle power of chest is closely related with respiratory function, although these muscles are not directly used for respiration.

[5] The prognosis of MND is generally not good, and most of the patients will die within ten years because of respiratory failure. The only way for survival is to use automatic ventilator[2].

Short-Term Effect Several interesting rules are discovered:

$$[Major\_Pectoralis = 2] \rightarrow [PaO_2 : D]$$
$$(Accuracy : 0.72, Coverage : 0.53, w = 3(months)),$$
$$[Biceps < 3] \rightarrow [PaO_2 : A2]$$
$$(Accuracy : 0.82, Coverage : 0.62, w = 3(months)).$$
$$[Biceps > 4] \rightarrow [PaO_2 : ZA]$$
$$(Accuracy : 0.88, Coverage : 0.72, w = 3(months)).$$

These rules suggest that if the power of muscles around chest is low, then respiratory function will decrease within one year and that if the power of muscles in arms is low, then respiratory function will decrease within a few years.

Long-Term Effect The following interesting rules are discovered:

$$[MajorPectoralis : A3] \wedge [Quadriceps : A3] \rightarrow [Pa0_2 : A3]$$
$$(Accuracy : 0.85, Coverage : 0.53, w = 1(year)),$$
$$[Gastro : A3] \rightarrow [PaO_2 : A3]$$
$$(Accuracy : 0.87, Coverage : 0.52, w = 1(year)).$$

These rules suggest that if the power of muscles of legs change very slowly, then respiratory function will decrease very slow. In summary, the system discovers that the power of muscles around chest and its chronological characteristics are very important to predict the respiratory function and how long it takes for a patient to reach respiratory failure.

6.2 Discovered Results in CVD

The above rule discovery system was applied to a clinical database on cerebrovascular diseases (CVD), which has 2610 records, described by 12 classes. Each record followed up at least 10 years and the averaged number of attributes are 2715. A list of w, $\{w\}$ was set to $\{$10 years, 5 years, 1 year, 3 months, 2 weeks$\}$ and thresholds, δ_α and δ_κ were set to 0.60 and 0.30,respectively. One of the most important problems in CVD is whether CVD patients will suffer from mental disorders or dementia and how long it takes each patient to reach the status of dementia.

Non-temporal Knowledge Concerning the database on CVD, several interesting rules are derived. The most interesting results are the following positive and negative rules for thalamus hemorrhage:

$$[Sex = Female] \wedge [Hemiparesis = Left] \wedge [LOC : positive] \rightarrow Thalamus$$
$$(accuracy : 0.62, coverage : 0.33),$$
$$[Risk : Hypertension] \wedge [Sensory = no] \rightarrow Putamen$$
$$(accuracy : 0.65, coverage : 0.43),$$

Interestingly, LOC(loss of consciousness) under the condition of $[Sex = Female] \wedge [Hemiparesis = Left]$ is an important factor to diagnose thalamic damage. In this domain, any strong correlations between these attributes and others, like MND, have not been found yet. It will be our future work to find what factor will be behind these rules. However, these rules do not include the relations between dementia and brain functions.

Short-Term Effect As short-term rules, the following interesting rules are discovered:

$$
\begin{aligned}
&[Gastro : A1] \wedge [Quadriceps : A1] \rightarrow [Dementia : A2] \\
&\quad (accuracy : 0.71, coverage : 0.31, w = 3(months)), \\
&[Gastro : D] \wedge [TA : D] \rightarrow [Dementia : A2] \\
&\quad (accuracy : 0.74, coverage : 0.32, w = 3(months)).
\end{aligned}
$$

These rules suggests that the rapid decrease of muscle power in the lower extremities are weakly related with the appearance of dementia. However, these knowledge has never been reported and further investigation is required for interpretation.

Long-Term Effect As long-term rules, the following interesting rules are discovered:

$$
\begin{aligned}
&[JointPosition : A3] \wedge [Quadriceps : A3] \rightarrow [Dementia : A3] \\
&\quad (accuracy : 0.61, coverage : 0.35, w = 1(year)), \\
&[Gastro : A3] \wedge [Vibration : A3] \rightarrow [Dementia : A3] \\
&\quad (accuracy : 0.87, coverage : 0.33, w = 1(year)).
\end{aligned}
$$

These rules suggests that combination of the decrease of muscle power in the lower extremities and the increase of sensory disturbance are weakly related with the appearance of dementia. However, these knowledge has neither been reported and further investigation is required for interpretation.

7 Discussion

7.1 Extended Moving Average

This paper introduces combination of extended moving average methods as first preprocessing, extraction of qualitative trend as second preprocessing and rule discovery. As discussed in Section 3 and 4, this approach is inspired by rule discovery in time series introduced by Das[4]. However, the main differences between Das's approach and our approach are the following.

1. For smoothing data, extended moving average method is introduced.
2. The system incorporates domain knowledge about a continous attribute to detect its qualitative trend.

3. Qualitative trend are calculated for each time-constant.
4. Rules are discovered with respect to not only associations between qualitative trends but also non-temporal associations.

Using these methods, the system discovered several interesting patterns in a clinical database of different time constant.

The disadvantage of this approach is that the program is not good at extracting periodical behavior of disease processes, or the recurrence of some diseases because the qualitative trends do not support the detection of cycles in temporal behavior of attributes. For these periodical processes, autoregressive function analysis is much more useful[6]. It will be our future work to extend our approach so that it can deal with periodicity more clearly.

7.2 Focusing Mechanism

One of the characteristics in medical reasoning is a focusing mechanism, which is used to select the final diagnosis from many candidates[14]. For example, in differential diagnosis of headache, more than 60 diseases will be checked by present history, physical examinations and laboratory examinations. In diagnostic procedures, a candidate is excluded if a symptom necessary to diagnose is not observed.

This style of reasoning consists of the following two kinds of reasoning processes: exclusive reasoning and inclusive reasoning. [6] The diagnostic procedure will proceed as follows: first, exclusive reasoning excludes a disease from candidates when a patient does not have a symptom which is necessary to diagnose that disease. Secondly, inclusive reasoning suspects a disease in the output of the exclusive process when a patient has symptoms specific to a disease. These two steps are modeled as usage of two kinds of rules, exclusive rules and inclusive rules(probabilistic rules), the former of which corresponds to exclusive reasoning and the latter of which corresponds to inclusive reasoning.

As shown in Section 1, probabilistic rules in this paper corresponds to inclusive reasoning with temporal attributes. That is, rules discovered in this approach support inclusive temporal reasoning. However, in order to support medical decision making, induction of exclusive rules with temporal attributes should be considered because exclusive reasoning also plays an important role in medical reasoning. One of the problems with induction of temporal exclusive rules is that the value of coverage may be changed with temporal evolution. Originally, the value of coverage of exclusive rules should be 1.0, so if the coverage of attributes has been changed, then it will not be included in the exclusive rules. One way to solve this problem is to relax the condition that the value of coverage is equal to 1.0. However, if the threshold of the

[6] Relations this diagnostic model with another diagnostic model are discussed in [15].

value of coverage is set, extended exclusive rules may be included into the category of probabilistic rules: we should define exclusive and inclusive rules in a more sophisticated way. It will be future work to develop induction of exclusive rules with temporal attributes.

8 Conclusion

In this paper, we introduce a combination of extended moving average method, multiscale matching and rule induction method, to discover new knowledge in temporal databases. In the system, extended moving average method are used for preprocessing, to deal with irregularity of each temporal data. Using several parameters for time-scaling with multiscale matching, given by users, this moving average method generates a new database for each time scale with summarized attributes. Then, rule induction method is applied to each new database with summarized attributes. This method was applied to two medical datasets, the results of which show that interesting knowledge is discovered from each database.

References

1. Abiteboul, S., Hull, R., and Vianu, V. *Foundations of Databases*, Addison-Wesley, New York, 1995.
2. Adams, R.D. and Victor, M. *Principles of Neurology*, 5th edition, McGraw-Hill, NY, 1993.
3. Agrawal, R., Imielinski, T., and Swami, A., Mining association rules between sets of items in large databases, in *Proceedings of the 1993 International Conference on Management of Data (SIGMOD 93)*, pp. 207-216, 1993.
4. Das, G., Lin, K.I., Mannila, H., Renganathan, G. and Smyth, P. Rule discovery from time series. In: *Proceedings of Fourth International Conference on Knowledge Discovery and Data Mining*, pp.16-22, 1998.
5. Fayyad, U.M., et al.(eds.)., *Advances in Knowledge Discovery and Data Mining*, AAAI Press, 1996.
6. Hamilton, J.D. *Time Series Analysis*, Princeton University Press, 1994.
7. Langley, P. *Elements of Machine Learning*, Morgan Kaufmann, CA, 1996.
8. Mokhtarian, F. and Mackworth, A. Scale-Based Description and Recognition of Planar Curves and Two-Dimensional Shapes. *IEEE Trans. Pattern. Anal. Machine Intell.*, **PAMI-8**, pp.34-43, 1986.
9. Miksch, S., Horn, W., Popow, C., and Paky, F. Utilizing temporal data abstraction for data validation and therapy planning for artificially ventilated newborn infants. *Artificial Intelligentce in Medicine*, **8**, 543-576, 1996.
10. Pawlak, Z., *Rough Sets*. Kluwer Academic Publishers, Dordrecht, 1991.
11. Rissanen J: *Stochastic Complexity in Statistical Inquiry*. World Scientific, Singapore, 1989.
12. Skowron, A. and Grzymala-Busse, J. From rough set theory to evidence theory. In: Yager, R., Fedrizzi, M. and Kacprzyk, J.(eds.) *Advances in the Dempster-Shafer Theory of Evidence*, pp.193-236, John Wiley & Sons, New York, 1994.

13. Tsumoto, S. and Tanaka, H., PRIMEROSE: Probabilistic Rule Induction Method based on Rough Sets and Resampling Methods. *Computational Intelligence*, **11**, 389-405, 1995.
14. Tsumoto S and Tanaka H: Automated Discovery of Medical Expert System Rules from Clinical Databases based on Rough Sets. In: *Proceedings of the Second International Conference on Knowledge Discovery and Data Mining 96*, AAAI Press, Palo Alto CA, pp.63-69, 1996.
15. Tsumoto, S., Automated Induction of Medical Expert System Rules from Clinical Databases based on Rough Set Theory. *Information Sciences* **112**, 67-84, 1998.
16. Tsumoto, S. Knowledge Discovery in Medical MultiDatabases: A Rough Set Approach, *Proceedings of PKDD99*(in this issue), 1999.
17. Tsumoto, S. Mining Positive and Negative Knowledge in Clinical Databases based on Rough Set Model. *Proceedings of PKDD2001.*
18. Van Bemmel,J. and Musen, M. A. *Handbook of Medical Informatics*, Springer-Verlag, New York, 1997.

Record linkage methods for multidatabase data mining

Vicenç Torra[1] and Josep Domingo-Ferrer[2]

[1] Institut d'Investigació en Intel·ligència Artificial - CSIC
Campus UAB s/n, 08193 Bellaterra (Catalonia, Spain)
e-mail: vtorra@iiia.csic.es, http://www.iiia.csic.es/~vtorra/

[2] Dept. of Computer Science and Mathematics, Universitat Rovira i Virgili,
Av. Països Catalans 26, 43007 Tarragona (Catalonia, Spain)
e-mail: jdomingo@etse.urv.es, http://www.etse.urv.es/recerca/crises/

Abstract. This chapter reviews record linkage techniques, useful to link records in two different data files corresponding to the same individual. Both probability-based and distance-based are presented and compared.

1 Introduction

Record linkage is one of the existing preprocessing techniques used for data cleaning for distributed and non-homogoneous databases. Such databases contain information about the same individuals described using the same variables that, frequently, do not match due to accidental distortion of the data[1]. Record linkage techniques are applied in such cases to find the records that correspond to the same individuals and to make databases consistent. Multi-database mining, that typically combines databases from different sources and, therefore, non-homogeneous also benefits from these tools.

This chapter describes existing mechanisms for re-identifying those pairs of records in two different data files corresponding to the same individual. We review in this chapter record linkage techniques in the case where the files share a set of variables. Techniques for files not sharing any variable have recently been proposed in [10] and will not be discussed here.

The structure of the chapter is as follows. Section 2 describes the notation that will be used in the rest of the paper. Then, in Sections 3 and 4 we review probabilistic and distance-based record linkage. Section 5 includes some technical issues concerning both record linkage approaches (*e.g.* standardization of variables and string comparison methods). The chapter ends with some conclusions.

---

[1] Sometimes differences are due to intentional distortion provoked *e.g.* for data anonymization

2 Notation

In this section we present the notation used throughout the chapter. We start describing the files and their records and, later on, the comparisons between pairs of records.

Let $\mathbf{A}$ and $\mathbf{B}$ be two data files defined, as usual, as sets of records. Let $r^{A,i}$ and $r^{B,i}$ denote the i-th record in files $\mathbf{A}$ and $\mathbf{B}$, respectively. We will use r^i when there is no possibility of confusion.

Records are defined in terms of variables (*i.e.* fields or attributes) and the values they take for those variables. Since files $\mathbf{F}$ contain a value for each record-variable pair, they can be modeled as a function:

$$\mathbf{V} : \mathbf{F} \rightarrow D(V_1) \times D(V_2) \times \cdots \times D(V_n)$$

where $\mathbf{V} = \{V_1, V_2, \cdots V_n\}$ denote the variables and $D(V_i)$ refer to the range of variable V_i (some fields like artificial intelligence often use the term domain of V_i). Without loss of generality, the n-dimensional function V can be assumed to be of the form:

$$\mathbf{V}(r) = (V_1(r)V_2(r) \cdots V_n(r))$$

where $V_i(\cdot) : \mathbf{F} \rightarrow D(V_i)$ is a one-dimensional function assigning a value for attribute V_i to a given record.

We shall use $\mathbf{V}^F(O)$ and V_i^F (for $i \in \{1, \ldots, n\}$) to specify the file when required.

For re-identification, we need to consider all pairs of records where one record is from the file $\mathbf{A}$ and the other is from the file $\mathbf{B}$. This is equivalent to considering the product of both files:

$$\mathbf{A} \times \mathbf{B}$$

Then, $r^{\mathbf{A}\times\mathbf{B}} = (a, b) \in \mathbf{A} \times \mathbf{B}$ denotes an arbitrary element of this set. Naturally, $a \in \mathbf{A}$ and $b \in \mathbf{B}$. We also assume the cardinality $|\mathbf{A} \times \mathbf{B}|$ to be N. Similar to the case of files, we will use $r^{\mathbf{A}\times\mathbf{B},i}$ to denote the i-th pair in $\mathbf{A} \times \mathbf{B}$ and use r^i when no possibility of confusion arises.

Two sets $\mathbf{M}$ and $\mathbf{U}$ are considered over the above Cartesian product, such that $\mathbf{M} \cup \mathbf{U} = \mathbf{A} \times \mathbf{B}$ and $\mathbf{M} \cap \mathbf{U} = \emptyset$). The first set $\mathbf{M}$ corresponds to the pairs such that both records (the one in $\mathbf{A}$ and the one in $\mathbf{B}$) correspond to the same individual; these are called the *matched pairs*. The second set $\mathbf{U}$ corresponds to the pairs such that both records correspond to different individuals; these are called *unmatched pairs*.

The procedure to link records of both files (*record linkage*) can be viewed as a classification of each record pair $(a, b) \in \mathbf{A} \times \mathbf{B}$ as either belonging to $\mathbf{M}$ or $\mathbf{U}$.

When re-identification is based on the assumption that both files share a set of variables, it is relevant to consider the comparison between the two records that define a pair. To do such comparisons, we assume that the number of variables in both files is n and that both files present variables in the same order (*i.e.* $V_i^A = V_i^B$ for $i \in \{1, \ldots, n\}$). In that case, the pair (a, b) can, alternatively, be expressed as:

$$((V_1^A(a), V_2^A(a), \ldots, V_n^A(a)), (V_1^A(b), V_2^A(b), \ldots, V_n^A(b)))$$

Let us define the following function over $\mathbf{A} \times \mathbf{B}$

$$\gamma(a, b) = (\gamma_1(a, b), \ldots, \gamma_n(a, b))$$

Given a pair (a, b), we define $\gamma_i(a, b)$ as 1 if $V_i(a) = V_i(b)$, and as 0 if $V_i(a) \neq V_i(b)$.

If we do not care about the specific pair of records but only about the values, we shall use the following notation:

$$\gamma = (\gamma_1, \ldots, \gamma_n)$$

Note that γ is a vector in $\{0, 1\}^n$ (actually it can be viewed as a coincidence vector).

The set of all γ coincidence vectors is denoted by Γ. The maximum cardinality of this latter set is 2^n. Thus $\Gamma = \{\gamma^1, \gamma^2, \ldots, \gamma^{2^n}\}$ where the coincidence vectors γ^j for $j = 1, \ldots, 2^n$ are n-dimensional vectors:

$$\gamma^j = (\gamma_1^j, \ldots, \gamma_n^j)$$

3 Probabilistic record linkage

The goal of record linkage is to establish whether pairs of records $(a, b) \in \mathbf{A} \times \mathbf{B}$ either belong to the set $\mathbf{M}$ or to the set $\mathbf{U}$. This is, whether both records a and b correspond to the same individual or to different individuals. Record linkage can be achieved by means of the so-called linkage or decision rules. These rules classify pairs as linked (placing them in $\mathbf{M}$) or non-linked (placing them in $\mathbf{U}$). Moreover, as the available information is sometimes not enough to discriminate between matched and unmatched pairs, some decision rules consider an additional classification alternative: clerical pairs. This is, a pair is classified as clerical when it cannot be automatically classified neither in $\mathbf{M}$ nor in $\mathbf{U}$; classification of clerical pairs must be manually done by a human operator. According to the above discussion, the following classes are considered by decision rules: $\mathbf{DR} = \{\mathbf{LP}, \mathbf{CP}, \mathbf{NP}\}$.

1. **LP**: Set of linked pairs
2. **CP**: Set of clerical pairs
3. **NP**: Set of non-linked pairs

$Name^{\mathbf{A}}$	$Surname^{\mathbf{A}}$	$Age^{\mathbf{A}}$	$Name^{\mathbf{B}}$	$Surname^{\mathbf{B}}$	$Age^{\mathbf{B}}$
Joan	Casanoves	19	Joan	Casanovas	19
Pere	Joan	17	Pere	Joan	17
J.M.	Casanovas	35	J.Manel	Casanovas	35
Juan	Garcia	53	Juan	Garcia	53
Ricardo	Garcia	14	Ricard	Garcia	14
Pere	Garcia	18	Pere	Garcia	82
Juan	Garcia	18	Juan	Garcia	18
Ricard	Tanaka	14	Ricard	Tanaka	18

Table 1. Files **A** and **B**

In probabilistic record linkage, a basic assumption is that files share a set of variables. Taking this into account, decision rules *rl* are defined as mappings from the comparison space (the space of all comparisons Γ) into probability distributions over **DR**. If $\gamma \in \Gamma$ then $rl(\gamma) = (\alpha_1, \alpha_2, \alpha_3)$ where $\alpha_1, \alpha_2, \alpha_3$ are, respectively, the membership probabilities for $\{\mathbf{LP}, \mathbf{CP}, \mathbf{NP}\}$. Naturally, $\alpha_1 + \alpha_2 + \alpha_3 = 1$ and $\alpha_i \geq 0$.

Example 1. Let us consider the files **A** and **B** in Table 1. Both files contain 8 records and 3 variables (Name, Surname and Age). For the sake of understandability, the files are defined so that records in the same row correspond to matched pairs and records in different rows correspond to unmatched pairs. The goal of record linkage in this example is to classify all possible pairs so that pairs with both records in the same row are classified as linked pairs and all the other pairs are classified as non-linked pairs.

To do so, we consider all pairs $(a, b) \in \mathbf{A} \times \mathbf{B}$. These pairs and the corresponding $\gamma(a, b)$ are displayed in Table 2. In this example, $\Gamma = \{\gamma^1 = 000, \gamma^2 = 001, \gamma^3 = 010, \gamma^4 = 011, \gamma^5 = 100, \gamma^6 = 101, \gamma^7 = 110, \gamma^8 = 111\}$. Note that the number of different coincidence vectors (8) is much less than the number of pairs in $\mathbf{A} \times \mathbf{B}$ (64). Yet, in record linkage, the classification of any pair (a, b) in Table 2 is solely based on its corresponding coincidence vector $\gamma(a, b)$.

Let us consider in the example below a rule that will be used later on. Rather than being expressed in terms of a probability distribution on **DR**, the rule directly assigns a class to each pair of records using the following expression:

$$P(\gamma = \gamma(a, b) | (a, b) \in \mathbf{M}) / P(\gamma = \gamma(a, b) | (a, b) \in \mathbf{U}) \tag{1}$$

Example 2. Let (a, b) be a pair of records in $\mathbf{A} \times \mathbf{B}$ and let (lt, ut) be two thresholds (lower and upper) in $\mathbb{R}$ such that $lt < ut$. Then a possible decision rule is:

1. If $R_p(a, b) \geq ut$ then (a, b) is a Linked Pair (**LP**)

2. If $R_p(a,b) \leq lt$ then (a,b) is a Non linked Pair (**NP**)
3. If $lt < R_p(a,b) < ut$ then (a,b) is a Clerical Pair (**CP**)

where the index $R_p(a,b)$ is defined in terms of the vector of coincidences $\gamma(a,b)$ as follows:

$$R_p(a,b) = R(\gamma(a,b)) = \log(\frac{P(\gamma(a,b) = \gamma(a',b')|(a',b') \in \mathbf{M})}{P(\gamma(a,b) = \gamma(a',b')|(a',b') \in \mathbf{U})}) \quad (2)$$

Remark that, in the above example, $R(\gamma)$ does not really use the values in records a and b but only their coincidences. Therefore, two pairs (a,b) and (c,d) such that $\gamma(a,b) = \gamma(c,d)$ are classified in the same way. The rationale of Expression 2 is made clear in the rest of this section and the use of log is explained in Section 3.2. Nevertheless, note that this rule associates large values of R to those pairs whose γ is such that $P(\gamma = \gamma(a',b')|(a',b') \in \mathbf{M})$ is large and $P(\gamma = \gamma(a',b')|(a',b') \in \mathbf{U})$ is small. Therefore, larger values are assigned to $R(\gamma)$ when the probability of finding the coincidence vector γ is larger in **M** than in **U**. Otherwise, small values of R are assigned to coincidence vectors with larger probabilities in **U** than in **M**.

In what follows, we will use m^i and u^i to denote the conditional probabilities of the coincidence vector γ^i:

$$m^i = P(\gamma^i = \gamma(a',b')|(a',b') \in \mathbf{M}) \quad (3)$$

$$u^i = P(\gamma^i = \gamma(a',b')|(a',b') \in \mathbf{U}) \quad (4)$$

Example 3. Table 3 gives the computation of R for all pairs of records in Table 2. Probabilities $m^i = P(\gamma^i = \gamma(a',b')|(a',b') \in \mathbf{M})$ and $u^i P(\gamma^i = \gamma(a',b')|(a',b') \in \mathbf{U})$ have been estimated by the proportion of elements in either **M** or **U** with such coincidence vector γ^i. The table gives coincidence vectors ordered (in decreasing order) according to their R values.

In general, for any decision rule *rl*, the following two probabilities are of interest:

$$P(\mathbf{LP}|\mathbf{U}) = \mu \quad (5)$$

$$P(\mathbf{NP}|\mathbf{M}) = \lambda \quad (6)$$

Note that the above are the probabilities that the rule causes an error. In particular, the first probability corresponds to the classification as a linked pair of a pair that is not a matched pair. This situation corresponds to the so-called *false linkage.* The second probability corresponds to the classification as a non-linked pair of a matched pair. This situation corresponds to the so-called *false unlinkage.*

Example 4. Let (lt, ut) be the lower and upper thresholds used in the decision rule of Example 2. Then, the probabilities $\mu = P(\mathbf{LP}|\mathbf{U})$ and $\lambda = P(\mathbf{NP}|\mathbf{M})$ for this decision rule are equal to:

$$\mu = \sum_{i:\log(m^i/u^i)>ut} u^i$$

$$\lambda = \sum_{i:\log(m^i/u^i)<lt} m^i$$

Assume $lt = 1.5$ and $ut = 2.5$. Using the data from Examples 1 and 3, gives the following values for μ and λ:

$$\mu = 0/56 + 1/56 = 1/56 = 0.0178$$

$$\lambda = 0 + 0 + 0 + 1/8 = 0.125$$

In addition to the two conditional probabilities above, another probability is also relevant in decision rules: the probability of classifying pairs of records into the set **CP**. As this latter set corresponds to pairs that should be further revised, the smaller the probability, the better. Therefore, it is clear that, given the set of all decision rules with the same probabilities $P(\mathbf{LP}|\mathbf{U})$ and $P(\mathbf{NP}|\mathbf{M})$, we are interested in finding the one (or ones) with the smallest probability of classifying a pair as **CP**.

To that end, Fellegi and Sunter [11] considered the following definitions.

Definition 1. Let rl be a decision rule in the space Γ and let μ and λ be the two values in the interval $(0,1)$ for its conditional probabilities $P(\mathbf{LP}|\mathbf{U})$ and $P(\mathbf{NP}|\mathbf{M})$ (Expressions 5 and 6). Then rl is a *rule with levels* μ *and* λ and is expressed by $rl(\mu, \lambda, \Gamma)$

Definition 2. Let $\mathbf{rl}$ be the set of all decision rules over Γ with levels μ and λ. Then $rl(\mu, \lambda, \Gamma)$ is the *optimal decision rule* if it satisfies:

$$P(\mathbf{CP}|rl) \leq P(\mathbf{CP}|rl')$$

for all $rl'(\mu, \lambda, \Gamma)$ in $\mathbf{rl}$.

In these definitions, it is assumed that μ and λ lead to a non-empty set of decision rules. It is said that μ and λ are admissible when they satisfy simultaneously Expressions 5 and 6 and when the set of decision rules is not empty. See [11] for details on the admissibility of μ and λ.

Fellegi and Sunter define an optimal decision rule based on Expression 1 and, as will be seen later, the rule is similar to the one we have given in Example 2. This optimal decision rule is defined below:

Definition 3. [11] Let μ and λ be an admissible pair of error levels and σ be a permutation of $\{1,\ldots,|\Gamma|\}$ such that $\sigma(j) < \sigma(k)$ if:

$$\frac{P(\gamma^{\sigma(j)} = \gamma(a',b')|(a',b') \in \mathbf{M})}{P(\gamma^{\sigma(j)} = \gamma(a',b')|(a',b') \in \mathbf{U})} > \frac{P(\gamma^{\sigma(k)} = \gamma(a',b')|(a',b') \in \mathbf{M})}{P(\gamma^{\sigma(k)} = \gamma(a',b')|(a',b') \in \mathbf{U})} \tag{7}$$

and let *limit* and *limit'* be the indexes such that:

$$\sum_{i=1,limit-1} u^{\sigma(i)} < \mu \leq \sum_{i=1,limit} u^{\sigma(i)} \tag{8}$$

$$\sum_{i=limit'+1,|\Gamma|} m^{\sigma(i)} < \lambda \leq \sum_{i=limit',|\Gamma|} m^{\sigma(i)} \tag{9}$$

where u^i and m^i correspond to the conditional probabilities in Expression 3 and 4.

Then, the optimal decision rule ODR_p for the pair (a,b) is a probability distribution $(\alpha_1, \alpha_2, \alpha_3)$ on $\{\mathbf{LP}, \mathbf{CP}, \mathbf{NP}\}$ defined by $ODR_p(a,b) = ODR(\gamma(a,b))$ with ODR defined as follows:

$$ODR(\gamma^{\sigma(i)}) = \begin{cases} (1,0,0) & \text{si } 1 \leq i \leq limit - 1 \\ (P_\mu, 1-P_\mu, 0) & \text{si } i = limit \\ (0,1,0) & \text{si } limit < i < limit' \\ (0, 1-P_\lambda, P_\lambda) & \text{si } i = limit' \\ (0,0,1) & \text{si } limit' + 1 \leq i \leq |\Gamma| \end{cases} \tag{10}$$

and where P_μ and P_λ are the solutions of the equations:

$$u^{\sigma(limit)} P_\mu = \mu - \sum_{i=1}^{limit-1} u^{\sigma(i)} \tag{11}$$

$$m^{\sigma(limit')} P_\lambda = \lambda - \sum_{i=limit'+1}^{|\Gamma|} m^{\sigma(i)} \tag{12}$$

This decision rule is optimal. This is established in the next theorem:

Theorem 1. *[11] The decision rule in Definition 3 is a best decision rule on Γ at the levels μ and λ.*

According to the procedure outlined above, the classification of a pair (a,b) requires: (i) computing the coincidence vector γ; (ii) determining the position of this γ vector in Γ once elements in Γ are ordered according to Expression 7 (otherwise put, finding i such that $\gamma^{\sigma(i)} = \gamma$) and (iii) computing the probability distribution over $\mathbf{DR}$ for this $\gamma^{\sigma(i)}$.

3.1 Alternative expressions for decision rules

In the particular case that μ and λ satisfy the following equations:

$$\mu = \sum_{i=1,limit} u^{\sigma(i)} \tag{13}$$

$$\lambda = \sum_{i=limit',N_\Gamma} m^{\sigma(i)} \tag{14}$$

the decision rule in Expression 10 can be simplifiedas:

$$SimpODR(\gamma^{\sigma(i)}) = \begin{cases} (1,0,0) \text{ if } 1 \leq i \leq limit \\ (0,1,0) \text{ if } limit < i < limit' \\ (0,0,1) \text{ if } limit' \leq i \leq |\Gamma| \end{cases} \tag{15}$$

This rule uses σ, $limit$ and $limit'$ as given in Definition 3, and is also optimal under the established conditions for μ and λ in Expressions 13 and 14.

Nevertheless, when Equations 13 and 14 do not hold, Rule 15 is not applicable and the previous definition with probability distributions is needed. To avoid the use of such probability distributions, that make practical applications more complex, we can classify as clerical pairs (assign them to class **CP**) those pairs that lead to a $\gamma^{\sigma(i)}$ with $i = limit$ or $i = limit'$. This is equivalent to using the following decision rule:

$$AltDR(\gamma^{\sigma(i)}) = \begin{cases} (1,0,0) \text{ if } 1 \leq i \leq limit - 1 \\ (0,1,0) \text{ if } limit \leq i \leq limit' \\ (0,0,1) \text{ if } limit' + 1 \leq i \leq |\Gamma| \end{cases} \tag{16}$$

Note that in this rule μ is used as an error bound, because the probability $P(\mathbf{LP}|\mathbf{U})$ of the new rule is smaller than the probability of the previous rule and, thus, smaller than μ. This is proven in the next proposition. The same applies for λ.

Proposition 1. *Let* $P_{\text{ODR}}(\mathbf{LP}|\mathbf{U})$ *and* $P_{\text{AltDR}}(\mathbf{LP}|\mathbf{U})$ *be the probabilities of the optimal decision rule* ODR *in Definition 3 and of the decision rule in Equation 16. Then,*

$$P_{\text{AltDR}}(\mathbf{LP}|\mathbf{U}) < P_{\text{ODR}}(\mathbf{LP}|\mathbf{U}) = \mu$$

Proof. Let us consider the probability $P_{ODR}(\mathbf{LP}|\mathbf{U})$ when the decision rule in Definition 3 is used for a given admissible pair of errors μ and λ. In this case, $P_{ODR}(\mathbf{LP}|\mathbf{U})$ equals to:

$$P_{ODR}(\mathbf{LP}|\mathbf{U}) = \sum_{i=1}^{limit-1} u^{\sigma(i)} + P_\mu \cdot u^{\sigma(limit)}$$

According to Equation 11 in Definition 3, $P_\mu \cdot u^{\sigma(limit)}$ equals to $\mu - \sum_{i=1}^{limit-1} u^{\sigma(i)}$. Therefore,

$$P_{ODR}(\mathbf{LP}|\mathbf{U}) = \sum_{i=1}^{limit-1} u^{\sigma(i)} + \mu - \sum_{i=1}^{limit-1} u^{\sigma(i)} = \mu$$

Alternatively, when the decision rule in Definition 16 is used, the following conditional probability is used:

$$P_{AltDR}(\mathbf{LP}|\mathbf{U}) = \sum_{i=1}^{limit-1} u^{\sigma(i)}$$

According to Equation 8, the above probability is less than μ. Therefore:

$$P_{AltDR}(\mathbf{LP}|\mathbf{U}) = \sum_{i=1}^{limit-1} u^{\sigma(i)} < \mu$$

Proposition 2. *Let* $P_{\mathrm{ODR}}(\mathbf{NP}|\mathbf{M})$ *and* $P_{\mathrm{AltDR}}(\mathbf{NP}|\mathbf{M})$ *be the probabilities of the optimal decision rule* ODR *in Definition 3 and of the decision rule in Equation 16. Then,*

$$P_{\mathrm{AltDR}}(\mathbf{NP}|\mathbf{M}) < P_{\mathrm{ODR}}(\mathbf{NP}|\mathbf{M}) = \lambda$$

Nevertheless, the rule in Expression 16 also classifies as clerical pairs those pairs (a, b) with $\gamma(a, b) = \gamma^{\sigma(limit)}$ or $\gamma(a, b) = \gamma^{\sigma(limit')}$ when Expressions 13 and 14 hold. Since these pairs can be classified as **LP** and **NP** without violating the bounds

$$P(\mathbf{LP}|\mathbf{U}) \leq \mu$$

and

$$P(\mathbf{NP}|\mathbf{M}) \leq \lambda$$

, the previous rule can be rewritten as follows:

$$rule(\gamma^{\sigma(i)}) = \begin{cases} (1,0,0) \text{ if } 1 \leq i \leq limit \\ (0,1,0) \text{ if } limit < i < limit' \\ (0,0,1) \text{ if } limit' \leq i \leq N_\Gamma \end{cases} \tag{17}$$

This requires that the indices *limit* and *limit'* be determined according to the following inequalities:

$$\sum_{i=1}^{limit} u^{\sigma(i)} \leq \mu < \sum_{i=1}^{limit+1} u^{\sigma(i)} \tag{18}$$

$$\sum_{i=limit'}^{|\Gamma|} m^{\sigma(i)} \leq \lambda < \sum_{i=limit'-1}^{|\Gamma|} m^{\sigma(i)} \tag{19}$$

It is important to underline that these latter rules are non-optimal rules because the probability of classifying a pair as a clerical pair is larger than the one in Definition 3. However, from a practical point of view, the last rule is convenient and easy to use; for example, it was the rule used in [14].

For the application of the decision rules defined so far, we need to know the position of the coincidence vector $\gamma^{\sigma(i)}$ in the ordering obtained from Γ and also the indexes *limit* and *limit'*. We give below an alternative definition that does not require these elements. This definition is equivalent to the one given above when appropriate thresholds are selected. This rule corresponds to the one presented in Example 2.

Definition 4. Let (a, b) be a pair of records in $\mathbf{A} \times \mathbf{B}$, let (lt, ut) be two thresholds (lower and upper) in $\mathbb{R}$ such that $lt < ut$, then the *Decision Rule* is defined as follows:

1. If $R_p(a, b) \geq ut$ then (a, b) is a Linked Pair (**LP**)
2. If $R_p(a, b) \leq lt$ then (a, b) is a Non linked Pair (**NP**)
3. If $lt < R_p(a, b) < ut$ then (a, b) is a Clerical Pair (**CP**)

where the index $R_p(a, b)$ is defined in terms of the vector of coincidences $\gamma(a, b)$ using Equation 1:

$$R_p(a, b) = R(\gamma(a, b)) = \log\left(\frac{P(\gamma(a, b) = \gamma(a', b')|(a', b') \in \mathbf{M})}{P(\gamma(a, b) = \gamma(a', b')|(a', b') \in \mathbf{U})}\right) \quad (20)$$

Proposition 3. *The decision rule in Definition 4 is equivalent to the decision rule in Expression 17 when lt and ut are defined as follows:*

$$ut = \log(m^{\sigma(limit)}/u^{\sigma(limit)})$$

$$lt = \log(m^{\sigma(limit')}/u^{\sigma(limit')})$$

where, as usual, $m^i = P(\gamma^i = \gamma(a', b')|(a', b') \in \mathbf{M})$, $u^i = P(\gamma^i = \gamma(a', b')|(a', b') \in \mathbf{U})$.

Example 5. Let us consider the probabilistic record linkage of records defined in terms of the three variables (*Name, Surname, Age*) as in Example 1. Let us consider the conditional probabilities m^i and u^i inferred from files **A** and **B** in Example 1 and computed in Example 3 (displayed in Table 3).

Now, let us compute the decision rule for $\mu = 0.05$ and $\lambda = 0.2$ and show its application to classify the pair *((J. Gomez 19), (P. Gomez 19))*.

First, to define the rule, we need to determine *limit* and *limit'* to apply Proposition 3. These values are set by Expressions 18 and 19 and the conditional probabilities in Table 3. Taking all this into account, we get

$$\sum_{i=1}^{2} u^{\sigma(i)} = 0 + 0.017 \leq 0.05 < 0 + 0.017 + 0.035 = \sum_{i=1}^{2+1} u^{\sigma(i)}$$

$$\sum_{i=5}^{|\Gamma|} m^{\sigma(i)} = 0+0+0+0+0.125 \leq 0.2 < 0+0+0+0+0.125+0.375 = \sum_{i=5-1}^{|\Gamma|} m^{\sigma(i)}$$

Therefore, $limit = 2$ and $limit' = 5$. Thus, $ut = \log(m^{\sigma(limit)}/u^{\sigma(limit)}) = \log(14) = 2.63$ and $lt = \log(m^{\sigma(limit')}/u^{\sigma(limit')}) = \log(3.5) = 1.25$.

According to this, the rule becomes:

1. If $R_p(a,b) \geq 2.63$ then (a,b) is a Linked Pair
2. If $R_p(a,b) \leq 1.25$ then (a,b) is a Non linked Pair
3. If $lt < R_p(a,b) < ut$ then (a,b) is a Clerical Pair

Now, we can consider any pair of records and classify them using this rule. If we take the pair $(J.Gomez19)$, $(P.Gomez19)$ we first compute the coincidence vector γ. We get $\gamma = (011)$. Then, we need to compute for this vector $R(011)$. Using the values m^i and u^i in Table 3 we get $R(011) = \log(14) = 2.63$. Therefore, the rule classifies the pair as a linked pair.

In this section we have seen how to define the decision rule and how to apply to a pair of records. However, this process requires several conditional probabilities to be determined. One possibility is to start with a pair of records for which the matched pairs are known (as in the examples in this section) and then estimate the probabilities by proportions of records. In the next sections we consider in more detail the computation of $R(a,b)$ and the estimation of the probabilities involved in $R(a,b)$.

3.2 Computation of $R_p(a,b)$

Some general aspects about the computation of $R_p(a,b)$ for a given pair (a,b) in $\mathbf{A} \times \mathbf{B}$ are described in this section. Specifically, the estimation of the probabilities involved in this computation is detailed in Section 3.3. In fact, according to the rule, the computation of $R_p(a,b)$ is solely based on the computation of R for $\gamma(a,b)$.

Due to the fact that the cardinality of Γ is typically quite large (recall that $|\Gamma| = 2^n$ where n is the number of variables), it is not appropriate to directly estimate the probabilities of all γ. To avoid this computation, it is usual to assume that the components of the vector $\gamma = (\gamma_1, \ldots, \gamma_n)$ are statistically independent. Under this assumption, the probabilities $P(\gamma = \gamma(a,b)|(a,b) \in \mathbf{M})$ and $P(\gamma = \gamma(a,b)|(a,b) \in \mathbf{U})$ can be expressed in the following way:

$$P(\gamma = \gamma(a,b)|(a,b) \in \mathbf{M}) = \prod_{i=1,n} P(\gamma_i = \gamma_i(a,b)|(a,b) \in \mathbf{M}) \quad (21)$$

$$P(\gamma = \gamma(a,b)|(a,b) \in \mathbf{U}) = \prod_{i=1,n} P(\gamma_i = \gamma_i(a,b)|(a,b) \in \mathbf{U}) \quad (22)$$

To simplify the notation, we shall use the following equivalences:

- $m(\gamma) = P(\gamma = \gamma(a,b)|(a,b) \in \mathbf{M})$
- $m_i(\gamma_i) = P(\gamma_i = \gamma_i(a,b)|(a,b) \in \mathbf{M})$
- $u(\gamma) = P(\gamma = \gamma(a,b)|(a,b) \in \mathbf{U})$
- $u_i(\gamma_i) = P(\gamma_i = \gamma_i(a,b)|(a,b) \in \mathbf{U})$

Using these equivalences, Equations 21 and 22 are rewritten as:

$$m(\gamma) = \prod_{i=1,n} m_i(\gamma_i)$$

$$u(\gamma) = \prod_{i=1,n} u_i(\gamma_i)$$

Therefore, under the same conditions of independence $R_p(a,b)$ can be rewritten as:

$$R_p(a,b) = R(\gamma(a,b)) = \log(\frac{P(\gamma(a,b) = \gamma(a',b')|(a',b') \in \mathbf{M})}{P(\gamma(a,b) = \gamma(a',b')|(a',b') \in \mathbf{U})}) \quad (23)$$

$$= \log(\frac{m(\gamma(a,b))}{u(\gamma(a,b))}) \quad (24)$$

$$= \log(\frac{\prod_{i=1,n} m_i(\gamma_i(a,b))}{\prod_{i=1,n} u_i(\gamma_i(a,b))}) \quad (25)$$

$$= \sum_{i=1,n} \log(m_i(\gamma_i(a,b))/u_i(\gamma_i(a,b))) \quad (26)$$

Note that the use of logarithm in $R(\gamma)$ simplifies its expression. Now, using that the following expressions about conditional probabilities hold for all $i \in \{1,\dots,n\}$

$$P(\gamma_i = 1|(a,b) \in \mathbf{M}) + P(\gamma_i = 0|(a,b) \in \mathbf{M}) = m_i(1) + m_i(0) = 1$$

$$P(\gamma_i = 1|(a,b) \in \mathbf{U}) + P(\gamma_i = 0|(a,b) \in \mathbf{U}) = u_i(1) + u_i(0) = 1$$

we define

- $m_i = m_i(1)$
- $u_i = u_i(1)$

and express $m_i(0)$ and $u_i(0)$ as:

$$m_i(0) = 1 - m_i$$

$$u_i(0) = 1 - u_i$$

These definitions permits to express the conditional probabilities in an alternative and more compact way (note that here γ_i and $1 - \gamma_i$ is either 1 or 0):

$$P(\gamma = \gamma(a',b')|(a',b') \in \mathbf{M}) = \prod m_i^{\gamma_i}(1 - m_i)^{1-\gamma_i}$$

$$P(\gamma = \gamma(a', b') | (a', b') \in \mathbf{U}) = \prod u_i^{\gamma_i} (1 - u_i)^{1-\gamma_i}$$

By further defining $w_i(\gamma_i)$ as $\log(m_i(\gamma_i)/u_i(\gamma_i))$, we have that $R(a, b)$ in Expression 26 can be rewritten as:

$$R_p(a, b) = R(\gamma(a, b)) = \sum_{i=1}^{n} w_i(\gamma_i(a, b)) \quad (27)$$

Thanks to the above definitions, we only need m^i and u^i to compute $w_i(\gamma_i(a, b))$. To do so, two cases are considered:

Case $\gamma_i(a, b) = 1$: define $w_i(1) = log(m_i/u_i)$.
Case $\gamma_i(a, b) = 0$: define $w_i(0) = log((1 - m_i)/(1 - u_i))$.

Note that these expressions are correct because $w_i(1)$ is equal to $\log(m_i(1)/u_i(1)$ and $m_i = m_i(1)$ and $u_i = u_i(0)$. Similary, $w_i(0)$ is equal to $\log(m_i(0)/u_i(0)$ and, thus, considering the equalities $m_i(0) = 1 - m_i$ and $u_i(0) = 1 - u_i$ we get the expression above.

The terms $w_i(\gamma_i(a, b))$ are known as the *weights* of $\gamma_i(a, b)$. As the usual case is to have $m_i > u_i$, then, the variables with coincident values (*i.e.* with $\gamma_i = 1$) contribute positively to the value $R(\gamma)$. Instead, variables with non-coincident values (i.e., with $\gamma_i = 0$) contribute negatively to the value $R(\gamma)$.

In fact, expressions for $w_i(1)$ and $w_i(0)$ given above clearly show that, when $m_i > u_i$, the weights for $\gamma_i(a, b) = 1$ are positive (and thus contribute positively to $R_p(a, b)$) and the weights for $\gamma_i(a, b) = 0$ are negative (and thus contribute negatively to $R_p(a, b)$). This is stated in the next proposition.

Proposition 4. *Let $m_i > u_i$, then $w_i(1) > 0$ and $w_i(0) < 0$.*

Proof. $m_i > u_i$ implies $m_i/u_i > 1$, therefore $w_i(1) = \log \frac{m_i}{u_i} > \log 1 = 0$. Also, it implies $1 - u_i > 1 - m_i > 0$, therefore, $1 > \frac{1-m_i}{1-u_i}$ and, thus, $0 = \log 1 > \log \frac{1-m_i}{1-u_i}$

Let us now turn into the estimation of probabilities m_i and u_i for all $i \in \{1, \ldots, n\}$.

3.3 Estimation of the probabilities

The estimation of the probabilities involved in the computation of R_p is usually based on the EM (Expectation-Maximization) algorithm [4]. In this section, we describe this method. We start reviewing the maximum likelihood model, then we describe the EM algorithm and we finish with its application to the record linkage process. The section finishes with the computation of the thresholds.

Likelihood function and maximum likelihood The maximum likelihood is a method for estimating the parameters of a given probability density. Let us consider a probability density $f(z|\theta)$. This is, f is a parametric model of the random variable z with parameter θ (or, parameters, because θ can be a vector). Let $\mathbf{z} = \{z_1, \ldots, z_e\}$ be a sample of the variable z. Then, the *likelihood* of z under a particular model $f(z|\theta)$ is expressed by:

$$f(\mathbf{z} = (z_1, \ldots, z_e)|\theta) = \prod_{i=1}^{e} f(z_i|\theta)$$

This is, $f(\mathbf{z}|\theta)$ is the probability of the sample $\mathbf{z}$ under the particular model $f(z_i|\theta)$ with a particular parameter θ. The likelihood function is the function above when the sample is taken as constant and θ is the variable. This is denoted by $L(\theta|\mathbf{z})$. Thus,

$$L(\theta|\mathbf{z}) = \prod_{i=1}^{e} f(z_i|\theta)$$

Often, the *log-likelihood* function is used instead of the likelihood function. The former is the logarithm of the latter and is denoted by $l(\theta|\mathbf{z})$ (or, sometimes, by $l(\theta)$). Therefore,

$$l(\theta|\mathbf{z}) = \log L(\theta|\mathbf{z}) = \log \prod_{i=1}^{e} f(z_i|\theta) = \sum_{i=1}^{e} \log f(z_i|\theta)$$

Given a sample $\mathbf{z}$ and a model $f(\mathbf{z}|\theta)$, the maximum likelihood estimate of the parameter θ is the $\hat{\theta}$ that maximizes $l(\theta|\mathbf{z})$. Equivalently, the estimate is $\hat{\theta}$ such that

$$l(\theta|\mathbf{z}) \leq l(\hat{\theta}|\mathbf{z})$$

EM algorithm The EM algorithm [4] (where EM stands for Expectation-Maximization) is an iterative process for the computation of maximum likelihood estimates. The method starts with an initial estimation of the parameters and then in a sequence of two step iterations builds more accurate estimations. The two steps considered are the so-called Expectation step and Maximization step.

The algorithm is based on the consideration of two sample spaces $\mathcal{Y}$ and $\mathcal{X}$ and a many-to-one mapping from $\mathcal{X}$ to $\mathcal{Y}$. We use y to denote this mapping, and $X(y)$ to denote the set $\{x|y = y(x)\}$. Only data y in $\mathcal{Y}$ are observed, and data x in $\mathcal{X}$ are only observed indirectly through y. Due to this, x are referred to as complete data and y as the observed data.

Let $f(x|\theta)$ be a family of sampling densities for x with parameter θ, it is clear that the corresponding family of sampling densities $g(y|\theta)$ can be computed from $f(x|\theta)$ as follows:

$$g(y|\theta) = \int_{X(y)} f(x|\theta)dx$$

Now, roughly speaking, the expectation step consists on estimating the complete data x and the maximization step consists on finding a new estimation of the parameters θ by maximum likelihood. In this way, the EM algorithm tries to find the value θ that maximizes $g(y|\theta)$ given an observation y. However, the method also uses $f(x|\theta)$.

EM algorithm for record linkage The application of EM to record linkage relies on the consideration of pairs of vectors $< \gamma(r), c(r) >_{r \in \mathbf{A} \times \mathbf{B}}$ as the complete data. Here, as usual, γ is the coincidence vector for $r \in \mathbf{A} \times \mathbf{B}$ and c (c for class) is a two dimensional vector $c = (c_m c_u)$ in $\{(10), (01)\}$ to indicate whether r belongs to $\mathbf{M}$ or $\mathbf{U}$. Then, for all pairs of records r in $\mathbf{A} \times \mathbf{B}$ we consider $(\gamma(r), c(r))$ where $c(r) = (10)$ if and only if $r \in \mathbf{M}$ and $c(r) = (01)$ if and only if $r \in \mathbf{U}$.

Incomplete data correspond to the case that some vectors c are unknown for some records r. Then, the expectation step assigns to the missing indicators fractions that sum to unity that are expectations given the current estimate of the parameters.

Here, the parameters θ consist of probabilities $m = (m_1, \ldots, m_n)$ and $u = (u_1, \ldots, u_n)$ with $m_i = P(1 = \gamma_i(a,b)|(a,b) \in \mathbf{M})$ and $u_i = P(1 = \gamma_i(a,b)|(a,b) \in \mathbf{U})$ (as defined in Section 3.2). Additionally, the parameters θ also contains p (the proportion of matched pairs $p = |M|/|M \cup U|$). Therefore, $\theta = (m, u, p)$.

Then, the log-likelihood for the complete data corresponds to [4]:

$$\ln f(\mathbf{x}|\theta) = \sum_{j=1}^{N} c(r^j)(\ln P\{\gamma(r^j)|\mathbf{M}\}, \ln P\{\gamma(r^j)|\mathbf{U}\})^T + \sum_{j=1}^{N} g(r^j)(\ln p, \ln(1-p))^T$$

This expression allows us to estimate the probabilities of assigning records in $\mathbf{A} \times \mathbf{B}$ either to the class $\mathbf{M}$ or $\mathbf{U}$ once an estimation for the parameters $\theta = (m, u, p)$ is given. In fact, the assignment does only depend on the corresponding coincidence vector. Therefore, the estimation is computed for the coincidence vectors $\gamma^j \in \Gamma$. This is the expectation step (see [14]), which yields the following assignment probabilities:

$$\hat{c}_m(\gamma^j) = \frac{\hat{p} \prod_{i=1}^{n} \hat{m}_i^{\gamma_i^j} (1 - \hat{m}_i)^{1-\gamma_i^j}}{\hat{p} \prod_{i=1}^{n} \hat{m}_i^{\gamma_i^j} (1 - \hat{m}_i)^{1-\gamma_i^j} + (1 - \hat{p}) \prod_{i=1}^{n} \hat{u}_i^{\gamma_i^j} (1 - \hat{u}_i)^{(1-\gamma_i^j)}} \quad (28)$$

$$\hat{c}_u(\gamma^j) = \frac{(1-\hat{p})\prod_{i=1}^{n}\hat{u}_i^{\gamma_i^j}(1-\hat{u}_i)^{1-\gamma_i^j}}{\hat{p}\prod_{i=1}^{n}\hat{m}_i^{\gamma_i^j}(1-\hat{m}_i)^{1-\gamma_i^j} + (1-\hat{p})\prod_{i=1}^{n}\hat{u}_i^{\gamma_i^j}(1-\hat{u}_i)^{(1-\gamma_i^j)}} \tag{29}$$

Then, in the maximization step, we need to calculate a new estimation of the parameters in θ. Therefore, we compute $\hat{m}_i$, $\hat{u}_i$ for all variables $i \in \{1, ..., n\}$ and recompute $\hat{p}$. This is done using the following equations (see [14]):

$$\hat{m}_i = \frac{\sum_{j=1}^{N}[\hat{c}_m(\gamma(r^j))\gamma_i(r^j)]}{\sum_{j=1}^{N}[\hat{c}_m(\gamma(r^j))]} \tag{30}$$

$$\hat{u}_i = \frac{\sum_{j=1}^{N}[\hat{c}_u(\gamma(r^j))\gamma_i(r^j)]}{\sum_{j=1}^{N}[\hat{c}_u(\gamma(r^j))]} \tag{31}$$

$$\hat{p} = \frac{\sum_{j=1}^{N}[\hat{c}_m(\gamma(r^j))]}{N} \tag{32}$$

Although the latter equations are written to consider all pairs of records in $\mathbf{A} \times \mathbf{B}$, it is advisable to accumulate the frequencies of each coincidence vector γ and use alternative expressions. If $fq(\gamma^j)$ is the frequency of the γ^j coincidence vector, then equations above for $\hat{m}_i$, $\hat{u}_i$ and $\hat{p}$ can be rewritten as:

$$\hat{m}_i = \frac{\sum_{j=1}^{2^n}[\hat{c}_m(\gamma^j)\gamma_i^j fq(\gamma^j)]}{\sum_{j=1}^{2^n}[\hat{c}_m(\gamma^j)fq(\gamma^j)]} \tag{33}$$

$$\hat{u}_i = \frac{\sum_{j=1}^{2^n}[\hat{c}_u(\gamma^j)\gamma_i^j fq(\gamma^j)]}{\sum_{j=1}^{2^n}[\hat{c}_u(\gamma^j)fq(\gamma^j)]} \tag{34}$$

$$\hat{p} = \frac{\sum_{j=1}^{2^n}[\hat{c}_m(\gamma^j)fq(\gamma^j)]}{\sum_{i=1}^{2^n} fq(\gamma^j)} \tag{35}$$

Initialization step In [14], it is stated that the algorithm is not very sensitive to initial values for m and u, although values $m_i > u_i$ are advisable. $m_i = 0.9$ was reported in [14] and $m_i = 0.9$ and $u_i = 0.1$ were used in [7] and [9].

4 Distance-based Record Linkage

This approach, described in [22] in a very restricted formulation for disclosure risk assessment, consists of computing distances between records in the two data files being considered. Then, the pair of records at minimum distance are considered linked pairs. We give below its formulation.

Let $d(a, b)$ be a distance function between records in file **A** and file **B**. Then, the distance-based record linkage is defined in the following way:

for all $a \in \mathbf{A}$
begin
  b' = arg $\min_{b \in \mathbf{B}} d(a, b)$
  $\mathbf{LP} = \mathbf{LP} \cup (a, b')$
  for all $b \in \mathbf{B}$ **such that** $b \neq b'$
  begin
    $\mathbf{NP} = \mathbf{NP} \cup (a, b)$
  end
end

Naturally, application of the approach relies on the existence of the distance function. Thus, a distance is assumed in each variable V_i. We denote this distance by d_{V_i}. The following distance has been considered:

Definition 5. Let $\mathbf{V} = \{V_1, V_2, \cdots V_n\}$ be the set of variables and let d_{V_i} be a distance on the range of $D(V_i)$. Then, assuming equal weight for all variables, the distance between records a and b is defined by:

$$d(a, b) = \sum_{i=1}^{n} d_{V_i}(V_i^A(a), V_i^B(b))$$

Several alternatives can be considered as within-variable distances d_V. In particular, depending on the type of variable, the following distances have been used [7], [9]:

Definition 6.
1. For a numerical variable V, the Euclidean distance d_E is used. Thus, $d_V = d_E$. In order to avoid scaling problems, it is convenient to standardize the Euclidean distance.
2. For a nominal variable V, the only permitted operation is comparison for equality. This leads to the following distance definition:

$$d_V(c, c') = \begin{cases} 0 \text{ if } c = c' \\ 1 \text{ if } c \neq c' \end{cases}$$

   where c and c' correspond to categories for variable V.
3. For an ordinal variable V, let $\leq_V$ be the total order operator over the range of V. Then, the distance between categories c and c' is defined as the number of categories between the minimum and the maximum of c and c' divided by the cardinality of the range:

$$d_V(c,c') = \frac{|c'' : min(c,c') \leq_V c'' \leq_V max(c,c')|}{|D(V)|}$$

4.1 Discussion

The performance of probabilistic and distance-based record linkage is compared using data from a public repository. The results of two experiments, one for numerical data and one for categorical data, are explained below. In short, a data file has been masked (*e.g.*, distorted) using some data protection mechanisms and then record linkage programs have been applied to the pair (original file, masked file). The number of correctly linked recods give a measure of record linkage performance.

Two data files were obtained from the U. S. Census Bureau Data Extraction System (DES [5]). One with numerical data and the other with categorical data. In the numerical case, we used records from the Current Population Survey (corresponding to 1995). In the categorical case, data from the American Housing Survey 1993 was used. Details on the variable and record selection are explained in [7].

Masking methods were applied to both files to obtain several masked files. Masking methods are applied by National Statistical Offices to perturb the original data so that data is protected and the anonymity of the original respondents is assured. Several masking methods with several parameterizations were applied to the original data and for each pair (method, parameterization) a perturbed data file was obtained. On the basis of the two original data files (*i.e.*, numerical and categorical) two groups of masked files can be distinguished.

The next step is to apply probabilistic record linkage and distance-based record linkage to each pair of masked file and original file. Tables 5 and 6 give the percentage of re-identified records using distance-based and probabilistic record linkage for the numerical data. Tables 8 and 9 give the percentage of re-identified records, again for distance-based and probabilistic record linkage, for categorical data. Tables 7 and 7 give the difference between distance-based and probabilistic record linkage.

The results obtained in these tables show that in the numerical case, distance-based record linkage outperforms probabilistic record linkage. In the categorical case, however, probabilistic and distance-based record linkage lead to quite similar results, being the former slightly better[2].

[2] These results complement the ones in [7] and [9] where high correlations between record linkage and several information loss measures were obtained. Therefore, both methods were considered similar for measuring disclosure risk

5 Technical issues

In this section we review several aspects that have to be taken in account when building record linkage implementations and when applying these methods. First, we consider the standardization process. This process is applied to names and addresses so that the probability of linking matched pairs increases. Then, we describe the use of blocking variables to reduce the number of pairs to be considered. The section continues with a review of some algorithms for string comparison and for partial agreement. The last part of the section is a discussion about the need of taking advantage of variables with non-uniform probability distributions.

5.1 Standardization of variables

To increase the performance of record linkage, standardization of some of the variables is recommended (in particular, variables corresponding to names, addresses and places). This process is required so that different forms of the same name (*e.g.*, Robert, Bob), company names (*e.g.*, Limited, Ltd., LTD) and addresses (*e.g.* street, st.) are transformed into a single form. If this process is not accomplished in an effective way, it is possible to classify as unlinked pairs some pairs that correspond to the same individual. Standardization consists of the three procedures below:

1. Parse variables to build a uniform structure
2. Detect relevant keywords to help in the process of recognizing the components that form the values of a variable
3. Replace all the (common) forms of a word by a single one (for example, an abreviation).

The goal of parsing is to ensure that, when the value of the variable consists of several elements, these always appear in the same order. For example "Robert Green, PhD", "Dr. Bob Green" and "Green, Robert" are translated into "PhD Robert Green", "Dr. Bob Green" and " Robert Green", respectively, following a *title + name + surname* structure.

The detection of special keywords can help in this process. For example, detection of "Ms", "PhD" or "Dr" is usually an indication of the presence of a personal name and "Ltd" indicates the presence of a company name. Detection would trigger specific parsing routines when appropriate.

The third standardization procedure replaces variants of values by a standard form. Depending on the meaning and the values of the variable, this procedure can either be applied to the whole variable value (*e.g.*, to the string used to represent the name) or to components of the variable value. This latter case occurs when the variable corresponds to personal names and they include for example title and middle letters, or when the variable is an address with street names, numbers or a P. O. box.

The substitutions required by standardization can be efficiently implemented by building a database with lists of words and their corresponding standard form so that the forms that appear in the files can be *replaced* by the standard ones. It is important to note that this *standard* form does not need to be a "dictionary" form (the root of a word or any not-shortened version of the name) but only an abstract identifier. This abstract identifier can be useful when a single spelling can have different origins (*e.g.* Bobbie might refer to Robert but also to Roberta).

For details on standardization see [28]. Examples of name and address parsing are provided there.

5.2 Blocking variables

When the files to be linked contain a huge number of records, consideration of all possible pairs is rather costly. This is so because $|\mathbf{A}| \cdot |\mathbf{B}|$ pairs have to be considered, where $\mathbf{A}$ and $\mathbf{B}$ are the files to be linked. Moreover, when each individual appears once in a file, only $\min(|\mathbf{A}|, |\mathbf{B}|)|$ pairs can be effectively linked. To avoid most of the unsuccessful comparisons, the so-called blocking variables (or blocking variables) are sometimes considered.

The set of blocking variables is selected by the user among the most error-resilient variables present in both files (those variables most likely to maintain their values across files). Given a set of blocking variables, comparison between pairs is restricted to those pairs with equal values for all blocking variables. In this way, the number of comparisons is largely reduced. To do such a comparison, files are usually ordered according to blocking variables. In this way, records to be compared are found in an easy way.

Naturally, when the blocking variables also contain errors, some of the linked pairs are not detected. These pairs are the so-called *missed matches*. Therefore, an unsuitable selection of the blocking variables results in a large number of missed matches.

A typical example of a blocking variable is the ZIP code. Nevertheless, it is also possible to use some string variables. In this case, a good alternative is to use the first letter or a particular coding so that all the symbols with a similar sound are mapped onto the same block (for example, the SOUNDEX codification – see Section 5.3).

To mitigate the negative effects of selecting a particular set of blocking variables (it is almost impossible to find error-free variables!), a good strategy is to apply several times the record linkage method using in each iteration a set of blocking variables independent from those used in previous iterations. Nevertheless, this process increases the complexity of the procedure. In any case, the use of blocking variables corresponds to a compromise between a high-cost detailed analysis of all possible pairs with few missed matches and a low-cost analysis of only a few pairs with more missed matches.

Blocking variables can also be used in combination with the EM algorithm. In fact, Jaro [14] used blocking variables when storing frequency counts in the EM algorithm in order to reduce computation. In such an approach, each file is partitioned into several blocks such that all records in a block have the same value for all the blocking variables. Then, pairs of records (one from each file) are only considered within the blocks with the same values for the blocking variables. For these pairs, coincidences are examined and counts are updated (the vector γ is computed and 1 is added to the frequency count for that particular configuration). Note that counters are not reset after a block is processed. In this way, counts represent the number of observations of each configuration over all blocks. However, it is important to note that, in this approach, counts are not the same that would be obtained without blocks. In fact, blocking reduces the number of unmatched pairs and, therefore, the probabilities u_i will be underestimated. To avoid underestimation, the probability

$$u_i = P(1 = \gamma_i(a,b)|(a,b) \in U)$$

is estimated by:

$$\hat{u}_i = P(1 = \gamma_i(a,b))$$

The above probability is the probability of $V_i(a) = V_i(b)$ and can be computed directly from the files **A** and **B** counting, for example, the number of pairs with the same value for the variable V_i:

$$\hat{u}_i = \frac{|\{V_i(a) = V_i(b)\}|}{N}$$

5.3 Partial coincidence and string comparison

In Section 3, we have described record linkage. We have only considered there the case of total coincidence between records (coincidence or non-coincidence). In practical situations, dealing with partial coincidence is also required. This topic is discussed in this section.

We start reviewing some algorithms for string comparison. First, we define the SOUNDEX method that transforms a string into a code that tends to bring together all variants of the same name. Therefore, the application of this method for string comparison leads to a Boolean comparison (strings are either encoded in the same or in a different way). Later on we review other string comparison methods that lead to values in the unit interval. The section finishes explaining how to adapt weights in probabilistic record linkage to accomodate partial coincidence.

SOUNDEX method A description of this method, originally developed by M. K. Odell and R. C. Russell [20,21], can be found in [15]. The method consists of transforming strings into a four character sequence. For example, both strings "Smith" and "Smythe" are encoded as "S530". Then, comparison between strings is achieved by means of comparison of sequences.

This coding has been used to deal with surnames. Jaro [14] recommends its use as a blocking variable:

> "To maximize the chance that similarly spelled surnames reside in the same block, the SOUNDEX system can be used to code the names, and the SOUNDEX code can be used as a blocking variable. There are better encoding schemes than SOUNDEX, but SOUNDEX with relatively few states and poor discrimination helps ensure that misspelled names receive the same code" (p. 418 in [14])

However, in the case of non-blocking variables, [14] does not recommend this coding because nonphonetic errors result in different codes, and, therefore, variants of the same name may receive different codes. The coding is said to be quite effective [18] except when the names are predominantly of Oriental origin.

Now we turn to the description of the method. As said above, the method transforms any string into a sequence of one character and three digits. The encoding rules are as follows:

1. The first letter of the string is selected and used as the first character of the codification.
2. Vowels A, E, I, O, U and letter Y are not encoded. Letters W and H are also ignored.
3. All the other letters are encoded as follows:

B, F, P, V	encoded as 1
C, G, J, K, Q, S, X, Z	encoded as 2
D, T	encoded as 3
L	encoded as 4
M, N	encoded as 5
R	encoded as 6

4. When the coding results into two or more adjacent codes with the same value only one code is kept. The others are removed. *E. g.,* "S22" is reduced to "S2" and "S221" to "S21".
5. All strings are encoded into a string with the following structure: `Letter, digit, digit, digit`. Additional elements are truncated and in case the string is too short, additional "0" are appended.

Table 11 displays some examples taken from Knuth [15]. Examples of pairs of surnames that do not lead to the samme codification include *(Rogers, Rodgers)* and *(Tchebysheff, Chebyshev).*

There exist other methods that proceed in a way similar to SOUNDEX by transforming a large number of strings into a single codification. These methods are classified in [24] as hashing techniques (a description of hash functions, very common in data structures, can be found *e.g.* in [1]). For example, Blair [3] builds the so-called r-letter abbreviations. This procedure transforms all strings s to r-letter strings removing $length(s) - r$ irrelevant characters. In this method, relevance of a character is computed in terms of relevance of letters (*e.g.*, "A" has relevance 5 and "B" relevance 1) and relevance of position (*e.g.* relevance of second position is larger than relevance of first position). Some example codings for 4-letter abbreviations are: *Euler* and *Ellery* are transformed to `ELER` and *Tchebysheff* and *Chebyshev* are transformed to `ESHE`. *Rogers* is translated to `OERS` and *Rodgers* can either be translated to `OERS` or `GERS` (letters "O" and "G" in *Rodgers* have the same importance but only one of them can be deleted).

Bigrams An alternative method for measuring string similarity is the one based on the comparison of the so-called bigrams. A bigram is defined as a pair of consecutive letters in a string. Therefore, the word `bigram` contains the following bigrams: `bi, ig, gr, ra, am`. The value of the function *simB* applied to two strings *s1* and *s2* is a value in the $[0, 1]$ interval corresponding to the number of bigrams in common divided by the mean value of bigrams in both strings:

$$simB(s1, s2) = \frac{|bigrams(s1) \cap bigrams(s2)|}{(|bigrams(s1)| + |bigrams(s2)|)/2}$$

where $bigrams(s)$ corresponds to the bigrams in string s.

Naturally, this function defines a similarity function and is equal to 1 when both strings are equal.

As said above, bigrams correspond to two consecutive characters. In fact, the literature also considers the general structure of n-grams (n consecutive characters in a string – a substring of length n). Similarity measures have been considered for n-grams with $n > 2$. In particular, there exists one such measures for the so-called trigrams (n-grams for $n = 3$).

Jaro algorithm The algorithm, introduced in [13], consists of the following steps when applied to strings $s1$ and $s2$:

1. Compute the length of the strings $s1$ and $s2$ ($strLen1 = length(s1)$, $strLen2 = length(s2)$).
2. Find the number of common characters. These characters are the ones that appear in both strings at a distance that is at most $minLen/2$ where $minLen = \min(strLen1, strLen2)$:

   $$common = \{c | c \in chars(s1) \cap chars(s2) \text{ and } pos(s1) - pos(s2) \leq minLen/2\}$$

3. Find the number of transpositions among the common characters. A transposition happens whenever a common character from one string does

not appear in the same position as the corresponding character from the other string. Let *trans* be the number of transpositions.

Then, the Jaro similarity is defined as follows:

$$jaro(s1, s2) = \frac{1}{3}(\frac{common}{strLen1} + \frac{common}{strLen2} + \frac{1}{2}\frac{trans}{common})$$

McLaughlin, Winkler and Lynch have studied this similarity and defined some enhancements which, when combined with record linkage, improve the performance of the latter. The enhancements and results are given in [23].

Dynamic programming methods Another approach for computing similarities between strings is based on dynamic programming. Below we describe the algorithm for computing the Levenshtein distance between two strings [16] (see also [2] or [24]). This distance is defined on any pair of strings (not necessarily of the same length) and gives the same weight (assumed to be 1 in the algorithm below) to insertions, deletions and substitutions:

```
for i = 0 to m do
begin
  d[i,0] = i;
end
for i = 0 to n do
begin
  d[0,i] = 0;
end
for j = 1 to n do
begin
  for i = 0 to m do
  begin
    if s1[i] == s2[j] then
      begin
        d[i,j] = d[i-1,j-1];
      end
    else
      begin
        d[i,j] = 1 + min(d[i-1,j], d[i,j-1], d[i-1,j-1]);
      end
    end if
  end
end
return d[m,n];
```

From the above distance, a similarity can be computed using a non-increasing function such that $f(0) = 1$. Improvements of this method exist so that the computation time and the working space requirement are reduced. See [24] for details.

Adapting weights w_i If partial coincidence is considered when comparing variable values, the weights attached to variables (w_i following the notation in Section 3.2) must be updated according to the coincidence. Usually, the update is proportional to the similarity (for example, multiplying the weight by the similarity – $w_i'(\gamma_i(a,b)) = w_i(\gamma_i(a,b)) \cdot similarity(V_i(a), V_i(b))$). Moreover, to improve the performance of the method, updating the similarity function by applying a particular transformation f is sometimes required. Therefore, an expression similar to the one below is used:

$$w_i'(\gamma_i(a,b)) = w_i(\gamma_i(a,b)) \cdot f(similarity(V_i(a), V_i(b)))$$

A particular example of the transformation function is the following one used in [23]:

$$f(x) = \begin{cases} x^{0.2435} & \text{if } x > 0.8 \\ 0.0 & \text{if } x \leq 0.8 \end{cases}$$

When a file with known coincidences is available, it is possible to learn these functions from the examples in that file.

Variables with values not following uniform probability distributions An important aspect to be considered when defining matching probabilities is that not all values in the range of a variable occur with the same probability. This is obvious in the case of names and surnames (*e.g.*, the probability of finding the Japanese surname Tanaka in Catalonia is very low). Newcombe [17] introduced a method to take into account the frequencies when computing the weights. The intuitive idea is that coincidence on surnames with a large frequency is less relevant than coincidence on less frequent surnames. Therefore, the probability of being a linked pair when a less frequent surname is detected is larger than when it is a high frequency surname.

6 Conclusions

In this chapter we have reviewed main record linkage techniques (probabilistic and distance-based ones) and we have compared their results for both numerical and categorical data. While distance-based record linkage seems to be more appropriate for numerical data, probabilistic based one seems more appropriate in the case of categorical data.

Acknowledgment

Partial support of the European Community under the contract "CASC" IST-2000-25069 and of the Spanish Ministry of Science and Technology under the project "STREAMOBILE" (TIC2001-0633-C03-01/02) is acknowledged.

References

1. Aho, A. V., Hopcroft, J. E., Ullman, J. D., (1988), Data Structures and Algorithms, Addison-Wesley, USA
2. Baeza-Yates, R., Ribeiro-Neto, B., (1999), *Modern Information Retrieval*, Addison-Wesley, England.
3. Blair, C. R., (1960), A Program for Correcting Spelling Errors, *Information and Control*, 3 60-67.
4. Dempster, A. P., Laird, N. M., Rubin, D. B., (1977), Maximum Likelihood From Incomplete Data Via the EM Algorithm, *Journal of the Royal Statistical Society*, 39, 1-38.
5. DES, (2002), Data Extraction System, U. S. Census Bureau, http://www.census.gov/DES/www/welcome.html
6. Domingo-Ferrer, J., Torra, V., (2001), Disclosure Control Methods and Information Loss for Microdata, 91-110, in *Confidentiality, Disclosure, and Data Access: Theory and Practical Applications for Statistical Agencies*, P. Doyle, J. I. Lane, J. J. M. Theeuwes, L. M. Zayatz (Eds.), Elsevier.
7. Domingo-Ferrer, J., Torra, V., (2001), A Quantitative Comparison of Disclosure Control Methods for Microdata, 111-133, in *Confidentiality, Disclosure, and Data Access: Theory and Practical Applications for Statistical Agencies*, P. Doyle, J. I. Lane, J. J. M. Theeuwes, L. M. Zayatz (Eds.), Elsevier.
8. Domingo, J., Torra, V., (2002), Aggregation techniques for statistical confidentiality, in *Aggregation operators: New trends and applications*, (R. Mesiar, T. Calvo, G. Mayor, eds.), Heidelberg: Physica-Verlag, pp. 261-271.
9. Domingo-Ferrer, J., Torra, V., (2002), Validating distance-based record linkage with probabilistic record linkage, *Lecture Notes in Computer Science*, vol. 2504, pp. 207-215, 2002.
10. Domingo-Ferrer, J., Torra, V., (2003), Disclosure risk assessment in statistical disclosure control via advanced record linkage, *Statistics and Computing* (to appear).
11. Fellegi, I. P., Sunter, A. B., (1969), A theory for record linkage, *Journal of the American Statistical Association*, 64:328, 1183-1210.
12. Gill, L., (2001), *Methods for Automatic Record Matching and Linking and Their Use in National Statistics*, National Statistics Methodology Series no. 25, London: Office for National Statistics.
13. Jaro, M. A., (1978), *UNIMATCH: A record linkage system: User's Manual*, U. S. Bureau of the Census, Washington DC.
14. Jaro, M. A., (1989), Advances in record-linkage methodology as applied to matching the 1985 Census of Tampa, Florida, *Journal of the American Statistical Association*, 84:406, 414-420.
15. Knuth, D. E., (1973), *The Art of Computer Programming Vol. 3: Sorting and Searching*, Reading, MA: Addison-Wesley.
16. Levenshtein, V. I., (1965), Binary codes capable of correcting deletions, insertions, and reversals, *Doklady Academii nauk SSSR*, 163:4 845–8 (in Russian) (also in *Cybernetics and Control Theory*, 10:8 (1966) 707-10).
17. Newcombe, H. B., Kennedy, J. M., Axford, S. J., James, A. P., (1959), Automatic linkage of vital records, *Science*, 130, 954-959.
18. Newcombe, H. B. (1967), Record linking: the design of efficient systems for linking records into individuals and family histories, *American Journal of Human Genetics*, 19:3, part I.

19. Newcombe, H. B., (1988), *Handbook of Record Linkage*, Oxford University Press.
20. Odell, M. K., Russell, R. C., (1918), U. S. Patents 1261167
21. Odell, M. K., Russell, R. C., (1922), U. S. Patents 1435663
22. Pagliuca, D., Seri, G., (1999), *Some Results of Individual Ranking Method on the System of Enterprise Accounts Annual Survey*, Esprit SDC Project, Deliverable MI-3/D2.
23. Porter, E. H., Winkler, W. E., (1997), Approximate string comparison and its effect on an advanced record linkage system, Report RR97/02, Statistical Research Division, U. S. Bureau of the Census, USA.
24. Stephen, G. A., (1994), *String Searching Algorithms*, World Scientific Publishing Co, Singapore.
25. Torra, V., (2000), Towards the re-identification of individuals in data files with common variables, *Proc. of the 14th European Conference on Artificial Intelligence (ECAI2000)*, Berlin, Germany.
26. Torra, V., (2000a), Re-identifying individuals using OWA operators, *Proc. of the 6th Int. Conference on Soft Computing*, Iizuka, Fukuoka, Japan.
27. Torra, V., (2000b), On the use of aggregation operators in Data Mining, submitted.
28. Winkler, W. E., (1993), Matching and record linkage, Report RR93/08, Statistical Research Division, U. S. Bureau of the Census, USA.
29. Winkler, W. E., Thibaudeau, Y., (1991), An application of the Fellegi-Sunter model of record linkage to the 1990 U. S. Decennial Census, Report, Statistical Research Division, U. S. Bureau of the Census, USA.

$Name^{\mathbf{A}}$	$Surname^{\mathbf{A}}$	$Age^{\mathbf{A}}$	$Name^{\mathbf{B}}$	$Surname^{\mathbf{B}}$	$Age^{\mathbf{B}}$	$\gamma(a,b)$	$\gamma(a,b)$
Joan	Casanoves	19	Joan	Casanovas	19	101	γ^6
Joan	Casanoves	19	Pere	Joan	18	000	γ^1
Joan	Casanoves	19	J.Manel	Casanovas	35	010	γ^3
Joan	Casanoves	19	Juan	Garcia	53	000	γ^1
Joan	Casanoves	19	Ricard	Garcia	14	000	γ^1
Joan	Casanoves	19	Pere	Garcia	82	000	γ^1
Joan	Casanoves	19	Juan	Garcia	18	000	γ^1
Joan	Casanoves	19	Ricard	Tanaka	18	000	γ^1
Pere	Joan	17	Joan	Casanovas	19	000	γ^1
Pere	Joan	17	Pere	Joan	18	110	γ^7
Pere	Joan	17	J.Manel	Casanovas	35	000	γ^1
Pere	Joan	17	Juan	Garcia	53	000	γ^1
Pere	Joan	17	Ricard	Garcia	14	000	γ^1
Pere	Joan	17	Pere	Garcia	82	100	γ^5
Pere	Joan	17	Juan	Garcia	18	000	γ^1
Pere	Joan	17	Ricard	Tanaka	18	000	γ^1
J.M.	Casanovas	35	Joan	Casanovas	19	010	γ^3
J.M.	Casanovas	35	Pere	Joan	18	000	γ^1
J.M.	Casanovas	35	J.Manel	Casanovas	35	011	γ^4
J.M.	Casanovas	35	Juan	Garcia	53	000	γ^1
J.M.	Casanovas	35	Ricard	Garcia	14	000	γ^1
J.M.	Casanovas	35	Pere	Garcia	82	000	γ^1
J.M.	Casanovas	35	Juan	Garcia	18	000	γ^1
J.M.	Casanovas	35	Ricard	Tanaka	18	000	γ^1
Juan	Garcia	53	Joan	Casanovas	19	000	γ^1
Juan	Garcia	53	Pere	Joan	18	000	γ^1
Juan	Garcia	53	J.Manel	Casanovas	35	000	γ^1
Juan	Garcia	53	Juan	Garcia	53	111	γ^8
Juan	Garcia	53	Ricard	Garcia	14	010	γ^3
Juan	Garcia	53	Pere	Garcia	82	010	γ^3
Juan	Garcia	53	Juan	Garcia	18	110	γ^7
Juan	Garcia	53	Ricard	Tanaka	18	000	γ^1
Ricardo	Garcia	14	Joan	Casanovas	19	000	γ^1
Ricardo	Garcia	14	Pere	Joan	18	000	γ^1
Ricardo	Garcia	14	J.Manel	Casanovas	35	000	γ^1
Ricardo	Garcia	14	Juan	Garcia	53	010	γ^3
Ricardo	Garcia	14	Ricard	Garcia	14	011	γ^4
Ricardo	Garcia	14	Pere	Garcia	82	010	γ^3
Ricardo	Garcia	14	Juan	Garcia	18	010	γ^3
Ricardo	Garcia	14	Ricard	Tanaka	18	000	γ^1
Pere	Garcia	18	Joan	Casanovas	19	000	γ^1
Pere	Garcia	18	Pere	Joan	18	101	γ^6
Pere	Garcia	18	J.Manel	Casanovas	35	000	γ^1
Pere	Garcia	18	Juan	Garcia	53	010	γ^3
Pere	Garcia	18	Ricard	Garcia	14	010	γ^3
Pere	Garcia	18	Pere	Garcia	82	110	γ^7
Pere	Garcia	18	Juan	Garcia	18	011	γ^4
Pere	Garcia	18	Ricard	Tanaka	18	001	γ^2
Juan	Garcia	18	Joan	Casanovas	19	000	γ^1
Juan	Garcia	18	Pere	Joan	18	001	γ^2
Juan	Garcia	18	J.Manel	Casanovas	35	000	γ^1
Juan	Garcia	18	Juan	Garcia	53	110	γ^7
Juan	Garcia	18	Ricard	Garcia	14	010	γ^3
Juan	Garcia	18	Pere	Garcia	82	010	γ^3
Juan	Garcia	18	Juan	Garcia	18	111	γ^8
Juan	Garcia	18	Ricard	Tanaka	18	001	γ^2
Ricard	Tanaka	14	Joan	Casanovas	19	000	γ^1
Ricard	Tanaka	14	Pere	Joan	17	000	γ^1
Ricard	Tanaka	14	J.Manel	Casanovas	35	000	γ^1
Ricard	Tanaka	14	Juan	Garcia	53	000	γ^1
Ricard	Tanaka	14	Ricard	Garcia	14	101	γ^6
Ricard	Tanaka	14	Pere	Garcia	82	000	γ^1
Ricard	Tanaka	14	Juan	Garcia	18	000	γ^1
Ricard	Tanaka	14	Ricard	Tanaka	18	110	γ^7

Table 2. Product space $\mathbf{A} \times \mathbf{B}$ and corresponding Γ vectors

$Name^{\mathbf{A}}$	$Surname^{\mathbf{A}}$	$Age^{\mathbf{A}}$	$Name^{\mathbf{B}}$	$Surname^{\mathbf{B}}$	$Age^{\mathbf{B}}$	γ^i		M/U	m^i	u^i	m^i/u^i	$\log(m^i/u^i)$
Juan	Garcia	53	Juan	Garcia	53	111	γ^8	M	2/8	0/56	∞	∞
Juan	Garcia	18	Juan	Garcia	18	111	γ^8	M				
Pere	Joan	17	Pere	Joan	18	110	γ^7	M	3/8	2/56	10.5	2.35
Juan	Garcia	53	Juan	Garcia	18	110	γ^7	U				
Pere	Garcia	18	Pere	Garcia	82	110	γ^7	M				
Juan	Garcia	18	Juan	Garcia	53	110	γ^7	U				
Ricard	Tanaka	14	Ricard	Tanaka	18	110	γ^7	M				
Joan	Casanoves	19	Joan	Casanovas	19	101	γ^6	M	1/8	2/56	3.5	1.25
Pere	Garcia	18	Pere	Joan	18	101	γ^6	U				
Ricard	Tanaka	14	Ricard	Garcia	14	101	γ^6	U				
Pere	Joan	17	Pere	Garcia	82	100	γ^5	U	0/8	1/56	0	$-\infty$
J.M.	Casanovas	35	J.Manel	Casanovas	35	011	γ^4	M	2/8	1/56	14	2.63
Ricardo	Garcia	14	Ricard	Garcia	14	011	γ^4	M				
Pere	Garcia	18	Juan	Garcia	18	011	γ^4	U				
Joan	Casanoves	19	J.Manel	Casanovas	35	010	γ^3	U	0/8	11/56	0	$-infty$
J.M.	Casanovas	35	Joan	Casanovas	19	010	γ^3	U				
Juan	Garcia	53	Ricard	Garcia	14	010	γ^3	U				
Juan	Garcia	53	Pere	Garcia	82	010	γ^3	U				
Ricardo	Garcia	14	Juan	Garcia	53	010	γ^3	U				
Ricardo	Garcia	14	Pere	Garcia	82	010	γ^3	U				
Ricardo	Garcia	14	Juan	Garcia	18	010	γ^3	U				
Pere	Garcia	18	Juan	Garcia	53	010	γ^3	U				
Pere	Garcia	18	Ricard	Garcia	14	010	γ^3	U				
Juan	Garcia	18	Ricard	Garcia	14	010	γ^3	U				
Juan	Garcia	18	Pere	Garcia	82	010	γ^3	U				
Pere	Garcia	18	Ricard	Tanaka	18	001	γ^2	U	0/8	3/56	0	$-\infty$
Juan	Garcia	18	Pere	Joan	18	001	γ^2	U				
Juan	Garcia	18	Ricard	Tanaka	18	001	γ^2	U				
Joan	Casanoves	19	Pere	Joan	18	000	γ^1	U	0/8	36/56	0	$-\infty$
Joan	Casanoves	19	Juan	Garcia	53	000	γ^1	U				
Joan	Casanoves	19	Ricard	Garcia	14	000	γ^1	U				
Joan	Casanoves	19	Pere	Garcia	82	000	γ^1	U				
Joan	Casanoves	19	Juan	Garcia	18	000	γ^1	U				
Joan	Casanoves	19	Ricard	Tanaka	18	000	γ^1	U				
Pere	Joan	17	Joan	Casanovas	19	000	γ^1	U				
Pere	Joan	17	J.Manel	Casanovas	35	000	γ^1	U				
Pere	Joan	17	Juan	Garcia	53	000	γ^1	U				
Pere	Joan	17	Ricard	Garcia	14	000	γ^1	U				
Pere	Joan	17	Juan	Garcia	18	000	γ^1	U				
Pere	Joan	17	Ricard	Tanaka	18	000	γ^1	U				
J.M.	Casanovas	35	Pere	Joan	18	000	γ^1	U				
J.M.	Casanovas	35	Juan	Garcia	53	000	γ^1	U				
J.M.	Casanovas	35	Ricard	Garcia	14	000	γ^1	U				
J.M.	Casanovas	35	Pere	Garcia	82	000	γ^1	U				
J.M.	Casanovas	35	Juan	Garcia	18	000	γ^1	U				
J.M.	Casanovas	35	Ricard	Tanaka	18	000	γ^1	U				
Juan	Garcia	53	Joan	Casanovas	19	000	γ^1	U				
Juan	Garcia	53	Pere	Joan	18	000	γ^1	U				
Juan	Garcia	53	J.Manel	Casanovas	35	000	γ^1	U				
Juan	Garcia	53	Ricard	Tanaka	18	000	γ^1	U				
Ricardo	Garcia	14	Joan	Casanovas	19	000	γ^1	U				
Ricardo	Garcia	14	Pere	Joan	18	000	γ^1	U				
Ricardo	Garcia	14	J.Manel	Casanovas	35	000	γ^1	U				
Ricardo	Garcia	14	Ricard	Tanaka	18	000	γ^1	U				
Pere	Garcia	18	Joan	Casanovas	19	000	γ^1	U				
Pere	Garcia	18	J.Manel	Casanovas	35	000	γ^1	U				
Juan	Garcia	18	Joan	Casanovas	19	000	γ^1	U				
Juan	Garcia	18	J.Manel	Casanovas	35	000	γ^1	U				
Ricard	Tanaka	14	Joan	Casanovas	19	000	γ^1	U				
Ricard	Tanaka	14	Pere	Joan	17	000	γ^1	U				
Ricard	Tanaka	14	J.Manel	Casanovas	35	000	γ^1	U				
Ricard	Tanaka	14	Juan	Garcia	53	000	γ^1	U				
Ricard	Tanaka	14	Pere	Garcia	82	000	γ^1	U				
Ricard	Tanaka	14	Juan	Garcia	18	000	γ^1	U				

Table 3. Product space $\mathbf{A} \times \mathbf{B}$ and corresponding Γ vectors

$Name^{\mathbf{A}}$	$Surname^{\mathbf{A}}$	$Age^{\mathbf{A}}$	$Name^{\mathbf{B}}$	$Surname^{\mathbf{B}}$	$Age^{\mathbf{B}}$	γ^i		**M/U**	$m^{\sigma(i)}$	$u^{\sigma(i)}$	$m^{\sigma(i)}/u^{\sigma(i)}$
Juan	Garcia	53	Juan	Garcia	53	111	$\gamma^\sigma(1)$	**M**	2/8	0/56	∞
...											
J.M.	Casanovas	35	J.Manel	Casanovas	35	011	$\gamma^\sigma(2)$	**M**	2/8	1/56	14
...											
Pere	Joan	17	Pere	Joan	18	110	$\gamma^\sigma(3)$	**M**	3/8	2/56	10.5
...											
Joan	Casanoves	19	Joan	Casanovas	19	101	$\gamma^\sigma(4)$	**M**	1/8	2/56	3.5
...											
Pere	Joan	17	Pere	Garcia	82	100	$\gamma^\sigma(5)$	**U**	0/8	1/56	0
Joan	Casanoves	19	J.Manel	Casanovas	35	010	$\gamma^\sigma(6)$	**U**	0/8	11/56	0
...											
Pere	Garcia	18	Ricard	Tanaka	18	001	$\gamma^\sigma(7)$	**U**	0/8	3/56	0
...											
Joan	Casanoves	19	Pere	Joan	18	000	$\gamma^\sigma(8)$	**U**	0/8	36/56	0
...											

Table 4. Product space $\mathbf{A} \times \mathbf{B}$ and corresponding Γ vectors

1.19	0.93	1.39	2.17	1.92	2.43	2.50	0.69	0.95	3.90	1.52	5.01	6.07	7.51	9.02
19.34	16.80	19.22	17.99	19.76	17.43	20.81	17.78	20.41	17.10	17.82	15.93	16.85	23.78	23.4
22.88	22.77	14.26	13.72	29.70	11.73	13.20	13.00	13.12	31.73	49.38	14.66	15.65	12.82	51.0
54.31	54.72	22.21	27.70	56.38	19.13	19.21	11.96	36.06	47.26	9.66	10.01	58.97	23.85	61.5
35.37	45.54	66.97	58.51	12.67	60.11	67.90	69.19	7.02	77.34	75.42	3.16	5.57	85.19	87.1
97.37	97.84	97.96	97.66	97.58	97.39	97.63	97.79	3.43	4.26	7.66	3.94	4.02	4.54	3.37
3.16	3.65	4.15	3.70	3.97	4.13	3.72	4.27	3.88	4.03	4.55	4.83	4.35	43.05	3.02
2.13	1.36	1.10	0.93											

Table 5. Average percentage of linked records using distance-based record linkage: numerical variables

0.15	0.08	0.11	0.12	0.07	0.25	0.25	0.09	0.09	0.38	0.20	0.52	0.85	1.08	2.79
4.70	13.60	3.44	3.44	6.67	5.45	4.15	3.35	13.90	2.08	3.98	2.00	2.37	18.29	22.75
16.69	22.78	1.88	1.38	29.06	1.14	1.20	1.22	0.99	36.92	27.29	0.50	4.66	0.85	33.04
33.70	37.41	22.39	29.03	42.00	6.93	6.24	3.52	39.76	57.47	2.34	0.97	56.84	24.48	60.69
46.98	56.22	64.79	65.28	2.90	66.56	67.63	66.35	1.90	71.32	71.85	0.77	1.26	74.13	73.03
74.07	74.07	74.40	75.28	75.99	78.96	79.78	88.06	0.62	0.67	1.52	0.69	0.62	0.25	0.50
0.61	0.44	0.67	0.53	0.34	0.46	0.66	0.38	0.41	0.42	0.52	0.38	0.38	64.88	0.48
0.25	0.29	0.15	0.22											

Table 6. Average percentage of linked records using probabilistic record linkage: numerical variables

1.045	0.847	1.283	2.050	1.852	2.183	2.249	0.595	0.860	3.517	1.323	4.497	5.225	6.425	6.227
14.643	3.204	15.780	14.550	13.095	11.984	16.653	14.431	6.508	15.026	13.836	13.929	14.484	5.489	0.741
6.190	-0.003	12.381	12.341	0.635	10.595	11.997	11.786	12.130	-5.185	22.090	14.153	10.992	11.971	17.989
20.608	17.315	-0.185	-1.336	14.378	12.196	12.963	8.439	-3.704	-10.215	7.315	9.048	2.130	-0.635	0.847
-11.614	-10.675	2.183	-6.772	9.775	-6.455	0.265	2.842	5.119	6.019	3.571	2.394	4.312	11.058	14.114
23.294	23.770	23.558	22.381	21.587	18.439	17.857	9.735	2.804	3.585	6.138	3.254	3.399	4.286	2.870
2.553	3.214	3.479	3.175	3.624	3.664	3.056	3.889	3.466	3.611	4.034	4.444	3.968	-21.826	2.540
1.878	1.071	0.952	0.701											

Table 7. Difference between the average percentage of linked records using distance-based and probabilistic record linkage: numerical variables

57.2	35.2	22.5	17.1	12.6	9.2	6.8	2.2	1.1	72.2	37.2	34	33.2	32.3	31.3
30.2	27.2	15.6	99.4	89.9	76.7	55.6	35.5	22.6	6.8	2.3	0.9	99.5	98.4	96.7
91.7	80.2	49.3	24.3	20.2	16.7	98.6	95.6	94.3	93.9	91.4	89.6	88	87.7	84.9
100	99.7	99.7	99.4	99	98.5	98.1	98	98.1	90.6	74.6	58.9	35.7	25.9	17.9
10.5	5.8	5	96.6	87.8	83.6	80.5	79.7	75.8	50.3	27.5	25.5	50.2	26.3	17.6
10.1	6.6	4.3	3.6	0.3	0.3	87.6	74.6	57	44.2	30.9	26.5	21.3	18.6	16.9
95.2	93.7	92.1	90.7	74	72.3	68.6	60.8	56.8	98.6	86.1	65.1	32.6	10.3	8.3
0.9	0.3	0.3	96.5	91.2	81.5	71.1	56.3	42.3	24	21.5	19.7	99.6	99.2	96.9
94.6	91.2	86.3	82.4	75.3	68.7	97.8	93.8	91.6	90.5	88.2	82.8	79.2	78	75.3
98.7	97	92.5	91.7	87.7	86.1	86.2	84.8	83.3	99.2	98.5	97.3	97.3	96.4	97.5
95.8	96	95.3	85.3	61.7	44.7	20	9.5	7.8	4.2	4.2	4.2	92.5	79.3	65.2
47.3	33.8	26.1	24.4	20.7	17.7	97	90.9	87.4	82.5	77.2	71.6	66.1	55.4	47.6

Table 8. Average percentage of linked records using distance-based record linkage: categorical variables

53	79.9	82.8	82.4	69.1	69	66.6	57	59.8	19.5	44.8	42.8	41	38.6	37.2
36.7	36.7	59.3	41.5	40.2	38	37.9	38.5	46.4	88.2	85.6	78.2	11.8	10.4	9.3
8.3	9.3	14.8	44.8	57.6	55.9	41.4	41.1	43.8	42.1	41.3	40.2	37.8	39.1	38.7
11.8	11.7	11.4	11.5	10.9	10.1	9.6	10.2	9.8	38.9	31.8	28	24.6	23.3	19
26.6	50.9	42.5	11.4	8.8	7.4	5.4	5.2	3.8	7.7	35.6	32.5	100.1	69.9	57.7
44.7	27.9	16.1	7.9	5.1	5.1	100.1	93.2	74.9	55.3	42.8	35.7	31.8	32.9	30.5
100.1	99.3	99.2	98.3	95.4	93.6	88.9	82.6	78	100.1	100.1	91.7	73	83.5	69.5
35.5	5.1	5.1	99.9	99.3	96.2	83.3	68.5	51.8	29.5	30	33.8	99.9	99.7	99.3
98.6	97.9	93.4	89.4	84.4	78.5	96.6	92.1	89.7	88.1	84.3	80.3	78.9	75.9	73.4
98.8	95.2	93.6	90.9	90.6	86.5	85.3	82.9	80.1	99.7	99.3	98	97.5	97.6	96.7
95.9	96.1	94.8	100.1	89.1	74.9	49.3	42.9	45.8	68.8	4.5	18.8	100.1	91.4	78.3
57.4	40.4	35	38.6	44	39.8	99.7	96.8	95	91.7	86.2	79.8	73	67.6	57.1

Table 9. Average percentage of linked records using probabilistic record linkage: categorical variables

4.2	-44.7	-60.3	-65.3	-56.5	-59.8	-59.8	-54.8	-58.7	52.7	-7.6	-8.8	-7.8	-6.3	-5.9
-6.5	-9.5	-43.7	57.9	49.7	38.7	17.7	-3	-23.8	-81.4	-83.3	-77.3	87.7	88	87.4
83.4	70.9	34.5	-20.5	-37.4	-39.2	57.2	54.5	50.5	51.8	50.1	49.4	50.2	48.6	46.2
88.2	88	88.3	87.9	88.1	88.4	88.5	87.8	88.3	51.7	42.8	30.9	11.1	2.6	-1.1
-16.1	-45.1	-37.5	85.2	79	76.2	75.1	74.5	72	42.6	-8.1	-7	-49.9	-43.6	-40.1
-34.6	-21.3	-11.8	-4.3	-4.8	-4.8	-12.5	-18.6	-17.9	-11.1	-11.9	-9.2	-10.5	-14.3	-13.6
-4.9	-5.6	-7.1	-7.6	-21.4	-21.3	-20.3	-21.8	-21.2	-1.5	-14	-26.6	-40.4	-73.2	-61.2
-34.6	-4.8	-4.8	-3.4	-8.1	-14.7	-12.2	-12.2	-9.5	-5.5	-8.5	-14.1	-0.3	-0.5	-2.4
-4	-6.7	-7.1	-7	-9.1	-9.8	1.2	1.7	1.9	2.4	3.9	2.5	0.3	2.1	1.9
-0.1	1.8	-1.1	0.8	-2.9	-0.4	0.9	1.9	3.2	-0.5	-0.8	-0.7	-0.2	-1.2	0.8
-0.1	-0.1	0.5	-14.8	-27.4	-30.2	-29.3	-33.4	-38	-64.6	-0.3	-14.6	-7.6	-12.1	-13.1
-10.1	-6.6	-8.9	-14.2	-23.3	-22.1	-2.7	-5.9	-7.6	-9.2	-9	-8.2	-6.9	-12.2	-9.5

Table 10. Difference between the average percentage of linked records using distance-based and probabilistic record linkage: categorical variables

Surnames		Coding
Euler	Ellery	E460
Gauss	Ghosh	G200
Hilbert	Heilbronn	H416
Knuth	Kant	K530
Lloyd	Ladd	L300
Lukasiewicz	Lissajous	L222

Table 11. SOUNDEX Codification

Part 3:

Model Building

Modelling data by the Choquet integral

Michel Grabisch*

LIP6, Université de Paris VI, Paris, France

Abstract. The chapter makes a survey of works dealing with the Choquet integral as a general non linear regression model. It is shown that its use is however limited to commensurate variables, as it is the case for example for multicriteria evaluation and multiattribute classification. A large part is devoted to the various methods of identifying parameters of the model, essentially quadratic programming and genetic algorithms. A new approach based on genetic algorithms is also described. Lastly, related works on classification and subjective evaluation are mentionned.

1 Introduction

It is well known that the main tool for finding dependencies among data is the linear regression model, which expresses one or several variables in term of a linear combination of the others. Linear regression is based on the least square principle, and has been studied at length, its statistical performance is well known. In real situations however, the linear assumption happens to be often rather far from the reality, and models with low accuracy are produced. For this reason, more general models that offer more flexibility are looked for, the price to pay being that they are often much more complex to use.

In this chapter, we are interested in using the Choquet integral [4] as a general non linear regression model. The Choquet integral is a generalization of the Lebesgue integral, defined with respect to a non classical measure, often called *fuzzy measure*, or *non-additive measure* or also *capacity*. When the underlying universe is finite, the Lebesgue integral reduces to a (convex) linear combination, hence can be assimilated to a particular class of regression models, where the coefficients are all positive and sum up to one. Hence, the Choquet integral offers a more general model, more precisely, as it will be seen below, it offers a set of (convex) linear models, each of them being defined in a polytope or simplex (possibly unbounded polyhedra).

The Choquet integral has been successfully applied many times in classification [38,19,24], decision making under uncertainty [32,2,33], multicriteria decision making [8,36,25] and also data modelling [37]. The main difficulty is to determine efficiently the $2^n - 2$ coefficients of the model. This exponential complexity limits the applicability, although solutions to reduce this number to a polynomial size exist.

* On leave from THALES, Corporate Research Laboratory, Domaine de Corbeville, Orsay, France

In this chapter, we make a quick overview of the main works along this line, and describe recent works we performed. First sections are devoted to introduce the necessary material. In all the chapter we consider a *single* regression problem, i.e. we want to model *one* variable y by some other variables $x_1, \ldots, x_n$.

2 The linear regression model

We recall briefly the linear regression model. Let y be a variable which we suppose we can explain or predict using a vector of variables $x = (x_1\ x_2\ \cdots\ x_n)^T$.

The general framework is rooted in estimation theory (where y is supposed to be unobservable, and thus has to be estimated), and based on the pioneering work of Gauss (see e.g.[1] for details). We consider x, y as random variables denoted X, Y. The assumption that y depends on x is that the distribution of $Y|X$ (posterior to observation) is different from the a priori distribution Y. Let us denote by $\hat{Y}(X)$ the estimated value of Y given an observation of X. The minimization of the variance of error $Y - \hat{Y}(X)$ leads to the unique solution $\hat{Y}(X) = E[Y|X]$, i.e. the conditional expectation. The linear hypothesis says that the estimated valued should be a linear expression of X, that is $\hat{Y}(X) = \alpha + \beta^T x$. Moreover, the expressions of α and β which minimize the variance of the error are given by:

$$\beta = \Gamma_{XY}\Gamma_Y^{-1} \tag{1}$$

$$\alpha = E[Y] - \beta E[X] \tag{2}$$

where Γ_{XY}, Γ_Y are the covariance matrices of X, Y and Y, i.e. $\Gamma_{XY} = E[(X - E[X])(Y - E[Y])^T]$, and $\Gamma_Y = E[(Y - E[Y])(Y - E[Y])^T]$. It turns out that the above linear model coincides with the (optimal) conditional expectation model when X, Y are gaussian.

The usual case of linear regression, where x, y are deterministic but the model is supposed to give only an approximation up to a random (model) error e is a particular case of the above linear estimation model:

$$y = \alpha + \beta^T x + e \tag{3}$$

Values of α, β which minimize the variance of the model error e follow directly from the above, when we have at disposal N data (realizations of x, y, denoted x^l, y^l, $l = 1, \ldots, N$), replacing expected values and covarianve matrices by the corresponding empirical expressions.

Usually, some quantities expressing the goodness of fit of the model to the data are computed.

3 Fuzzy measures and the Choquet integral

We present briefly necessary concepts around fuzzy measures and the Choquet integrals. Comprehensive treatments of this topic can be found in [5,18,17,31,39].

Let $N = \{1, \ldots, n\}$ be a finite set. A *capacity* or *fuzzy measure*, *non-additive measure* on N is any set function $\mu : \mathcal{P}(N) \longrightarrow \mathbb{R}^+$ such that $\mu(\emptyset) = 0$ and $A \subset B \subset N$ implies $\mu(A) \leq \mu(B)$ (monotonicity). μ is said to be *non-monotonic* if monotonicity does not hold.

The *conjugate* fuzzy measure of μ, denoted $\bar{\mu}$, is defined by $\bar{\mu}(A) = \mu(N) - \mu(A^c)$.

The Möbius transform of μ, denoted m^μ or m if there is no fear of ambiguity, is a set function on N defined by

$$m^\mu(A) := \sum_{B \subset A} (-1)^{|A|-|B|} \mu(B), \quad \forall A \subset N. \tag{4}$$

A fuzzy measure is said to be *additive* if $\mu(A \cup B) = \mu(A) + \mu(B)$ whenever $A \cap B = \emptyset$. A *k-additive measure* is a fuzzy measure such that $m(A) = 0$ for all $A \subset N$ such that $|A| > k$, and there is at least one A of k elements such that $m(A) \neq 0$. 1-additivity coincides with additivity.

An important notion for the interpretation of fuzzy measures is the one of Shapley and intéraction indices. For any $i \in N$, the Shapley index [34] of i is defined by:

$$\phi_i := \sum_{K \subset N \setminus i} \frac{(n - |K| - 1)!|K|!}{n!} [\mu(K \cup \{i\}) - \mu(K)]. \tag{5}$$

The Shapley index satisfies $\sum_{i=1}^n \phi_i = \mu(N)$, and can be interpreted as the overall importance of i. The concept of interaction for a pair of elements $i, j \in N$ has been proposed by Murofushi and Soneda [28].

$$I_{ij} := \sum_{K \subset N \setminus \{i,j\}} \frac{(n - |K| - 2)!|K|!}{(n-1)!} [\mu(K \cup \{i,j\}) - \mu(K \cup \{i\}) - \mu(K \cup \{j\}) + \mu(K)]. \tag{6}$$

It represents a kind of (positive or negative) synergy between elements. The definition has been extended by the author to any number of elements [11]:

$$I(A) := \sum_{K \subset N \setminus A} \frac{(n - |K| - |A|)!|K|!}{(n - |A| + 1)!} \sum_{B \subset A} (-1)^{|A|-|B|} \mu(K \cup B), \forall A \subset N. \tag{7}$$

Note that $I(\{i\}) = \phi_i$, and $I(\{i, j\}) = I_{ij}$. Also, it is easy to show that for an additive measure, $I(A) = 0$ whenever $|A| > 1$, and $\phi_i = \mu(\{i\})$. More generally, for a k-additive measure, $I(A) = 0$ whenever $|A| > k$.

We introduce now the (discrete) Choquet integral on N. We assimilate (positive) real-valued functions on N to points in $\mathbb{R}^n_+$. Let $x \in \mathbb{R}^n_+$, with components $x_i, i = 1, \ldots, n$, and μ be a fuzzy measure. The *Choquet integral* of x w.r.t. μ is defined by:

$$\mathcal{C}_\mu(x) := \sum_{i=1}^n x_{\sigma(i)} [\mu(A_{\sigma(i)}) - \mu(A_{\sigma(i+1)})] \tag{8}$$

where σ is a permutation of the elements of N such that $x_{\sigma(1)} \leq \cdots \leq x_{\sigma(n)}$, $A_{\sigma(i)} := \{\sigma(i), \sigma(i+1), \ldots, \sigma(n)\}$, and $A_{\sigma(n+1)} := \emptyset$. Remark that for a given permutation σ, the region $\{x \in \mathbb{R}^n_+ | x_{\sigma(1)} \leq \cdots \leq x_{\sigma(n)}\}$ is a simplex. We call *canonical simplexes* the set of all such simplexes, considering all possible permutations on N. They form a partition of $\mathbb{R}^n_+$. Moreover, continuity is ensured over all canonical simplexes.

4 The Choquet integral regression model

The regression based on Choquet integral is a generalization of Eq. (3) in the following sense:

$$y = \alpha + \mathcal{C}_\mu(x) + e$$

using previous notations, where μ is a (non monotonic in general) fuzzy measure. From Section 3, we know that more explicitly the model writes:

$$y = \alpha + \sum_{i=1}^{n} w_{\sigma(i)} x_{\sigma(i)} + e, \tag{9}$$

with $w_{\sigma(i)} = \mu(A_{\sigma(i)}) - \mu(A_{\sigma(i+1)})$. Observe that in any case $\sum_{i=1}^n w_{\sigma(i)} = \mu(N) = 1$, but the $w_{\sigma(i)}$'s are positive (i.e. we get a convex sum) iff μ is monotone.

Although this model is clearly more general than the linear one, some restrictions have to be pointed out. The first one is that the definition of the Choquet integral (8) is given for positive integrands, hence x should belong to $\mathbb{R}^n_+$. For real-valued integrands, two definitions exist, the symmetric and the asymmetric ones, which coincide when the fuzzy measure is additive, i.e. when the Choquet integral model collapses to the linear one. As a consequence, when $x \in \mathbb{R}^n$, two models are possible. The expressions of the symmetric Choquet integral $\check{\mathcal{C}}_\mu$ (called also Šipoš integral) and asymmetric Choquet integral $\mathcal{C}_\mu$ are:

$$\begin{aligned}\check{\mathcal{C}}_\mu(x) &= \mathcal{C}_\mu(x^+) - \mathcal{C}_\mu(x^-)\\ \mathcal{C}_\mu(x) &= \mathcal{C}_\mu(x^+) - \mathcal{C}_{\overline{\mu}}(x^-)\end{aligned}$$

where $x_i^+ = x_i \vee 0$, and $x_i^- = -x_i \vee 0$, for all i.

The second point is more of importance, and it is rooted in the definition of the Choquet integral. Considering a permutation σ on N, the corresponding canonical simplex is the locus of the points satisfying $x_{\sigma(1)} \leq x_{\sigma(2)} \cdots \leq x_{\sigma(n)}$. These simplexes are central in the definition of the Choquet integral, but they suppose that one can meaningfully compare the variables x_i's ! In real applications, variables are most often expressed with different units, e.g. in medicine one try to model the blood pressure in terms of age, height and weight. Clearly, these three variables are not commensurate, so applying the Choquet integral here is meaningless.

To the opinion of the author, the last above mentionned point prevents the Choquet integral to be used as a general non linear regression model, in any situation, despite the fact that this kind of model is advocated by some authors as Wang *et al.* [42,41]. However, this drawback disappears if one considers only commensurate variables. This is always the case in multicriteria evaluation problems, and multi-attribute classification, a domain where the Choquet integral has been successfully applied many times (see Section 6 for some examples).

In multicriteria evaluation, N is the set of criteria or attributes, and x_i is not the value taken by attribute i for some object, but represents the *satisfaction degree* or *attractiveness* felt by the decision maker in view of the value taken by attribute i. Depending on the application and the precise meaning attached to the x_i's, the underlying scale could be bounded unipolar (e.g. $[0,1]$), unipolar (e.g. $\mathbb{R}^+$) or bipolar (e.g. $\mathbb{R}$). In this last case, the scale could be a difference scale or a ratio scale, which determines the kind of integral to be used (symmetric or asymmetric): see details in [16].

In classification, N is the set of attributes, and x_i represents the membership degree of a given object to a given class, knowing only the value of attribute i. The membership degree is a bounded unipolar concept, so that the underlying scale is $[0,1]$.

In these two domains of application, the fuzzy measure is asked to be monotone, since this entails the monotonicity of the model, a natural requirement in these two fields.

We end this section by giving some words on the advantage of such models. Coefficients in a linear regression model are easy to interpret, as they represent the "weight" of a given variable in the model. In the case of the Choquet integral model, there are too many weights, and these weights live only in some canonical simplex, so that at first sight, the model is not easily interpretable. However, an interpretation in terms of the Shapley value and interaction indices permits to have a clear view of the model. In fact, the merit of the Choquet integral is to bring a powerful tool to model interaction between variables (see an explanation of interaction in e.g. [8,10], see also a particularly simple geometrical interpretation when $n = 2$ in [13]), which is theoretically well founded (see an axiomatization in the spirit of the Shapley value in [21]).

5 Determining the coefficients of the model

In the case of the linear regression, a unique solution was obtained, based on results of estimation theory. In the case of Choquet integral, no such result is available, and the coefficients of the model (i.e. the fuzzy measure) are obtained through an optimization procedure, whose solution is not unique in general. As in the linear regression model, one wants to minimize the squared error, that is:

$$E = \sum_{l=1}^{N} [y^l - \mathcal{C}_\mu(x^l)]^2 \tag{10}$$

under constraints if the fuzzy measure is asked to be monotone. If no constraint (apart positiveness) exists, then the problem is not so much difficult to solve, since it reduces to a usual least square problem. Indeed, it can be shown (see e.g. [18,20], and originally [27]) that E can be expressed in a quadratic form:

$$\frac{1}{2}\mathbf{u}^T\mathbf{D}\mathbf{u} + \mathbf{c}^T\mathbf{u}$$

where $\mathbf{u}$ is a $(2^n - 2)$ dimensional vector containing all the coefficients of the fuzzy measure μ (except $\mu(\emptyset)$ and $\mu(N)$ which are fixed), $\mathbf{D}$ is a $(2^n - 2)$ dimensional square matrix, and $\mathbf{c}$ a $(2^n - 2)$ dimensional vector. The constraints of monotonicity can be expressed with $\mathbf{u}$ under a linear form:

$$\mathbf{A}\mathbf{u} + \mathbf{b} \geq \mathbf{0}$$

where $\mathbf{A}$ is a $n(2^{n-1} - 1) \times (2^n - 2)$ matrix, and $\mathbf{b}$ a $n(2^{n-1} - 1)$ dimensional vector. Thus we obtain a quadratic program, which can be solved using standard techniques. In [26], Miranda and Grabisch study at length the properties of this quadratic program. There is no unique solution in general, and one important question is to know the minimum number of data required in order to have a "good" solution. This point remains however not completely clear.

Although the preceding approach provides an optimal solution, it happens that in some cases (large n, few data,...) the problem becomes ill-conditioned and bad results occur. Also, in practical applications, the optimal solution obtained does not always satisfy the decision maker, giving "extreme" values (near 0 and 1) and far from the equilibrium point $\mu(A) = 1/|A|$ for all $A \subset N$. For this reason, the author has proposed a suboptimal algorithm, called HLMS (Heuristic Least Mean Squares), based on the gradient algorithm and the idea of equilibrium point [7]. This algorithm gives an error very near the optimal one, while being much less memory and time consuming.

A third way to solve the optimization problem is to use heuristic algorithms, such as genetic algorithms (GA). Many authors have proposed methods based on GA to determine the fuzzy measure, although most of them are restricted to λ-measures (see e.g. [3]). A typical approach to determine (general) fuzzy measures is the one presented by Wang et al. in [40]. We describe in the whole the method, and propose some improvement we performed.

- the encoding of μ is done as follows: $\mu(A)$ is coded in a gene, for all $A \subset N$, $A \neq \emptyset, N$, so that a chromosome has $2^n - 2$ genes coding a given μ. Each gene is coded as a binary number, using p bits.
- the population of chromosomes is between 100 and 1000, and is randomly generated (uniform numbers generated on $[0, 1]$, plus a test for monotonicity).

- the fitness of a chromosome is defined by $\frac{1}{1+E}$, where E is the above defined quadratic error.
- the probability of choosing parents is proportional to the fitness value. Then reproduction is done according to three different processes: two-point crossover, three-bit mutation, and two-point realignment (see [40] for details).
- the population size is kept constant by always selecting the best individuals by their fitness.

Although this is not explained in the paper, we suppose that a test of monotonicity is performed over new chromosomes, and those which do not satisfy the conditions are eliminated.

We proposed a more general version, allowing to handle k-additive measures, and improving the optimization process in order to reduce the learning time and improve performance.

1. **General features:** the algorithm accepts several error criteria (sum of squares, of absolute values, etc.), and several fuzzy integrals (Choquet, Šipoš Sugeno, etc.).
2. **Coding:** it is the same than Wang's approach, at the difference that we code the Möbius transform when dealing with k-additive measures. In this last case, the number of genes to code is only $\sum_{l=1}^{k}\binom{n}{l} - 1$.
3. **Selecting the genes:** recall that for a given datum x is associated a permutation σ_x on N ordering the components of x in increasing order, and to each permutation σ is associated a maximal chain C_σ in the lattice of values of μ: only the coefficients belonging to the chain are used in the computation of the integral. Henceforth, for a given set of learning data $\mathcal{X}$, the coefficients that will be effectively used in the computations form the set $\bigcup_{x\in\mathcal{X}} C_{\sigma_x}$, which may be much smaller than the whole lattice. Such genes are called *significant*. In the learning stage, only significant genes will be modified by reproduction; however due to monotonicity constraints, non significant genes may have to be updated (see below). Note that this procedure works only for the case where μ is encoded, not its Möbius transform.
4. **Reproduction, test of monotonicity and selection:** these operations are performed on the current population, unless some stagnation is observed, in which case the former population of best individuals is chosen.
   - There are several modes of reproduction which can be chosen, either producing two children or one child, using one point crossover or two points crossover, selecting parents either at random, or neighbours, or according to a probability depending on the fitness function, and using mutation or not.
   - the test of monotonicity depends on the type of information which is coded (either μ or its Möbius transform). In the latter, the standard test of monotonicity for the Möbius transform is performed. In the

former case, recall that only genes in $\bigcup_{x\in\mathcal{X}} C_{\sigma_x}$ (called *significant*) are modified. Non significant genes are nevertheless updated in order to satisfy monotonicity of μ, in a spirit similar as in [7]. Specifically, in the lattice formed by μ, any non significant gene $\mu(A)$ is updated by

$$\mu(A)^{\text{new}} = \max(\mu(A)^{\text{old}}, \bigvee_{i\in A} \mu(A \setminus i)).$$

The test of monotonicity for significant genes is to verify that their value is at least as large as all its lower neighbours, i.e. $\mu(A) \geq \mu(A\setminus i)$, $\forall i \in A$.
- the selection is done in the usual way, keeping the best individuals at constant population size.

5. **Initial population:** random generation of monotone fuzzy measures or Möbius transforms.

We have compared our approach with the algorithm of Wang on a small example ($n = 4$) given in [40]. Wang *et al.* found an error of 0.0088 in 2 minutes on a 120 MHz processor, while we found an error of 0.0001 in less than 5 seconds on a 733 MHz processor. This shows clearly the improvement.

Tests comparing on various data sets the GA approach with the above described quadratic programming approach show that, as expected, the latter outperforms the former, in precision (both can give very close results when n is low), but mainly in learning time. In our opinion, GA's are not useful for the Choquet integral, since classical optimal methods can do the job very efficiently, but they become useful for highly non linear functionals as the Sugeno integral. A careful study devoted to the Sugeno integral has yet to be done. We mention also a study due to Nettleton and Torra comparing GA's and classical optimization for OWA-like aggregation operators [29].

6 Related works and examples

This section gathers various related works, using Choquet integral as a basic tool for modelling.

6.1 Classification

The Choquet integral has been used as an information fusion tool for performing multi-sensor classification, each sensor being able to give a classification. The first attempt in this direction was done by Tahani and Keller [38], using a λ-measure. At the same time, Grabisch and Sugeno proposed a more general approach for classification, similar to the Bayesian one [22,23].

Basically, the idea is to describe each class by a set of typical fuzzy sets or possibility distribution (one per attribute), and to define a fuzzy measure per class, on the set of attributes. For class j, the fuzzy measure μ_j expresses to what degree coalitions of attributes are able to distinguish class j from the others. The method has been described at length in several publications, we refer the reader to the following literature: [19,9,12]. See also [18,24] for an overview of related works in classification and image processing.

6.2 Subjective evaluation

As said in Section 4, multicriteria evaluation is particularly suited to the use of a Choquet integral model (or similar integrals as the Sugeno integral. A precise distinction of these two types of integrals is however still a topic of research). The term "subjective" refers to the fact that most of concepts modelled so far are highly subjective, so that no clear mathematical model can be derived simply from analysis of the problem, as the following examples will made it clear. It is worth noting that the first application of fuzzy measure was indeed evaluation, of woman faces (how subjective!), by Sugeno in his thesis [35].

Many applications in this field have been done in Japan during the eighties, such as modelling opinion on nuclear energy [30], evaluating the quality of printed color images, speakers, and so on (see an overview of these applications in [18]), rather in an ad hoc way. Later, using the concept of interaction as a basic tool for analysis, and k-additive measures, the author has performed several applications, as the evaluation of richness of a cosmetic [14], of discomfort in a seated position during a long time [15], of mental work load when performing some task, etc.

6.3 The model of Kwon and Sugeno

In [36,25], Kwon and Sugeno propose a model of multicriteria evaluation based on Choquet integral, but in a rather different approach as the one we presented above. Considering a set N of n criteria, their argument is that, except for low values of n, a model based on fuzzy measures becomes intractable due to the exponential complexity (2^n coefficients). Based on works of Fujimoto [6] about inclusion-exclusion coverings, they propose to replace the Choquet integral defined with respect to a fuzzy measure μ on N by a sum of p Choquet integrals w.r.t. fuzzy measures $\mu_1, \ldots, \mu_p$ on subsets $C_1, \ldots, C_p$ of N, such that $\bigcup_{i=1}^{p} C_i = N$, i.e. $\{C_1, \ldots, C_p\}$ is a covering of N. In equation:

$$y = \sum_{i=1}^{p} \mathcal{C}_{\mu_i}(x) + e. \tag{11}$$

In the above, x in $\mathcal{C}_{\mu_i}(x)$ is of course restricted to C_i. The μ_i's are non monotonic fuzzy measures, and e is a modelling error supposed to be a zero mean Gaussian random variable with variance σ^2 (denoted $\mathcal{N}(0, \sigma^2)$). x and y represents satisfaction degrees, and are supposed to be commensurable.

The identification of the parameters of the model is done as follows. Considering N independent realizations of y with data $x^1, \ldots, x^N$, denoted $y^1, \ldots, y^N$, the joint distribution of $y^1, \ldots, y^N$ is $\prod_{l=1}^{N} \mathcal{N}(\sum_{i=1}^{p} \mathcal{C}_{\mu_i}(x^l), \sigma^2)$, and assuming that the covering $\{C_1, \ldots, C_p\}$ is known, the μ_i's are determined in order to minimize the residual variance σ^2 of the error, which amounts to the squared error criterion presented above:

$$\hat{\sigma}^2 = \frac{1}{N} \sum_{l=1}^{N} \left[y^l - \sum_{i=1}^{p} \mathcal{C}_{\mu_i}(x^l) \right]$$

Since the μ_i's are not restricted to be monotone, the estimation can be done with the least square method.

The second step is to determine the optimal covering. Here genetic algorithms are used. A particular covering is coded in a chromosome containing $2^n - 1$ genes corresponding to all subsets of N, except the empty set. The gene corresponding to a particular subset is set to 1 if the covering contains this subset, and to 0 otherwise. We refer the reader to [36,25] for further details of implementation, and detail only the fitness function. They use the Bayesian Information Criterion (BIC) defined as:

$$\text{BIC} = N \log \hat{\sigma}^2 + k \log N$$

where k is the number of independent parameters in the model. The criterion expresses the fact that a balance should be kept between precision of the model (low $\hat{\sigma}^2$) and simplicity (low k).

We make some comments about this model.

- Although equation (11) seems to be very general, it coincides exactly with a Choquet integral whenever the covering is an inclusion-exclusion covering (IEC) (see [6]). If not, it is more general, but properties of this class of operators have not been studied by the authors. The main interest of this way of decomposing the integral has the advantage of exhibiting some structure in the evaluation model and to lower complexity. Another way to do this is to use k-additive measures. In fact, Fujimoto and Murofushi have shown that an IEC corresponds more or less to the subsets where the Möbius transform is non zero (more exactly, the finest irreducible (i.e. non redundant) IEC is the set of all maximal subsets of non-zero Möbius transform).
- The use of non-monotonic measures, although more general, is questionable since we are here in an evaluation process, where x is a vector of satisfaction degree. It is a fundamental assumption in such context that the model is monotonic, i.e. the increase of a satisfaction degree on some criterion cannot decrease the overall satisfaction degree y. But this condition implies the monotonicity of the fuzzy measure.

- Results presented in [25] on real data (evaluation of motorcycles, according to different categories of population) show that the performance of the model are very close to a usual linear model (almost same error, sometimes better for the linear model, but with a slightly better BIC value). A much better modelling error is obtained using a (general) Choquet integral, at the price of having much more coefficients. Also, the subsets in the covering are almost reduced to singletons and sometimes pairs, which explains why the performance is so close to a linear model. This may be due to the fitness function since it is linear in k and only logarithmic in $\hat{\sigma}^2$.

6.4 Discovering links betweeen variables

Lastly we present briefly a general model of regression proposed by Wang *et al.* [41]. Supposing to have at disposal data about some variables, the algorithm tries to find the best dependencies among the variables, and each sub-model is given by a Choquet integral regression model.

Specifically, dependencies among variables are expressed as an oriented acyclic graph. Each node y with entering arrows can be explained by the parent nodes $x_1, \ldots, x_n$ (corresponding to the entering arrows), and this forms a sub-model. Each sub-model is expressed through a Choquet integral:

$$y = q\mathcal{C}_\mu + e$$

where q is a real constant (positive?), μ a fuzzy measure, and e is the modelling error, supposed to be Gaussian and centered on a value c.

The basic idea is the following: the best acyclic graph is obtained by genetic algorithms, while for each sub-model, an algorithm similar to the one proposed in [7] is used, and q, c are estimated as in a linear regression model.

To the opinion of the author, the approach, although very general and powerful, does not avoid the restriction we have indicated in Section 4. First of all, one cannot combine with a Choquet integral variables which are not commensurate, hence the above model cannot be used in data mining in general as claimed by the authors. Second, the fact that fuzzy measures are used (and not non monotonic fuzzy measures) implies that each sub-model is monotonically increasing w.r.t each variable, an assumption which is again restrictive.

7 Conclusion

In this chapter, we have presented an overview of methods using the Choquet integral as a tool of modelling data. The main advantage compared to linear models is that they are able to take into account interactions and dependency between variables. We have also presented the limitations of such approaches, which to our sense, should be limited to the modelling of commensurate variables, a situation which is typically encountered in multicriteria evaluation, multi-attribute classification (after all, a Choquet integral is not more than a generalization of expected value...).

We have also presented a new approach of determination of fuzzy measures by genetic algorithms. This could be used for other fuzzy integrals, such as the Sugeno integral, where classical optimization methods fail.

Acknowledgment

The author is grateful to Ch. de Rivière and J. Wasong for their active participation in the project of determining fuzzy measures by genetic algorithms.

References

1. B.D.O. Anderson and J.B. Moore. *Optimal filtering.* Prentice Hall, 1979.
2. A. Chateauneuf. Modeling attitudes towards uncertainty and risk through the use of Choquet integral. *Annals of Operations Research*, 52:3–20, 1994.
3. T.Y. Chen and J.C. Wang. Identification of λ-measures using sampling design and genetic algorithms. *Fuzzy Sets and Systems*, 123:321–341, 2001.
4. G. Choquet. Theory of capacities. *Annales de l'Institut Fourier*, 5:131–295, 1953.
5. D. Denneberg. *Non-Additive Measure and Integral.* Kluwer Academic, 1994.
6. K. Fujimoto and T. Murofushi. Hierarchical decomposition of the Choquet integral. In M. Grabisch, T. Murofushi, and M. Sugeno, editors, *Fuzzy Measures and Integrals — Theory and Applications*, pages 94–104. Physica Verlag, 2000.
7. M. Grabisch. A new algorithm for identifying fuzzy measures and its application to pattern recognition. In *Int. Joint Conf. of the 4th IEEE Int. Conf. on Fuzzy Systems and the 2nd Int. Fuzzy Engineering Symposium*, pages 145–150, Yokohama, Japan, March 1995.
8. M. Grabisch. The application of fuzzy integrals in multicriteria decision making. *European J. of Operational Research*, 89:445–456, 1996.
9. M. Grabisch. The representation of importance and interaction of features by fuzzy measures. *Pattern Recognition Letters*, 17:567–575, 1996.
10. M. Grabisch. Alternative representations of discrete fuzzy measures for decision making. *Int. J. of Uncertainty, Fuzziness, and Knowledge Based Systems*, 5:587–607, 1997.
11. M. Grabisch. k-order additive discrete fuzzy measures and their representation. *Fuzzy Sets and Systems*, 92:167–189, 1997.
12. M. Grabisch. Fuzzy integral for classification and feature extraction. In M. Grabisch, T. Murofushi, and M. Sugeno, editors, *Fuzzy Measures and Integrals — Theory and Applications*, pages 415–434. Physica Verlag, 2000.
13. M. Grabisch. A graphical interpretation of the Choquet integral. *IEEE Tr. on Fuzzy Systems*, 8:627–631, 2000.
14. M. Grabisch, J.M. Baret, and M. Larnicol. Analysis of interaction between criteria by fuzzy measure and its application to cosmetics. In *Int. Conf. on Methods and Applications of Multicriteria Decision Making*, pages 22–25, Mons, Belgium, May 1997.

15. M. Grabisch, J. Duchêne, F. Lino, and P. Perny. Subjective evaluation of discomfort in sitting position. *Fuzzy Optimization and Decision Making*, 1:287–312. 2002.
16. M. Grabisch and Ch. Labreuche. To be symmetric or asymmetric? A dilemna in decision making. In J. Fodor, B. De Baets, and P. Perny, editors, *Preferences and Decisions under Incomplete Knowledge*, pages 179–194. Physica Verlag, 2000.
17. M. Grabisch, T. Murofushi, and M. Sugeno. *Fuzzy Measures and Integrals. Theory and Applications (edited volume)*. Studies in Fuzziness. Physica Verlag, 2000.
18. M. Grabisch, H.T. Nguyen, and E.A. Walker. *Fundamentals of Uncertainty Calculi, with Applications to Fuzzy Inference*. Kluwer Academic, 1995.
19. M. Grabisch and J.M. Nicolas. On the performance of classification techniques based on fuzzy integrals. In *5th Int. Fuzzy Systems Assoc. Congress*, pages 163–166, Seoul, Korea, July 1993.
20. M. Grabisch and J.M. Nicolas. Classification by fuzzy integral — performance and tests. *Fuzzy Sets & Systems, Special Issue on Pattern Recognition*, 65:255–271, 1994.
21. M. Grabisch and M. Roubens. An axiomatic approach to the concept of interaction among players in cooperative games. *Int. Journal of Game Theory*, 28:547–565, 1999.
22. M. Grabisch and M. Sugeno. Fuzzy integral with respect to dual measures and its application to multi-attribute pattern recognition. In *6th Fuzzy Systems Symposium*, pages 205–209, Tokyo, Japan, September 1990. in japanese.
23. M. Grabisch and M. Sugeno. Multi-attribute classification using fuzzy integral. In *1st IEEE Int. Conf. on Fuzzy Systems*, pages 47–54, San Diego, CA, March 1992.
24. J.M. Keller, P.D. Gader, and A.K. Hocaoğlu. Fuzzy integrals in image processing and recognition. In M. Grabisch, T. Murofushi, and M. Sugeno, editors, *Fuzzy Measures and Integrals — Theory and Applications*, pages 435–466. Physica Verlag, 2000.
25. S.H. Kwon and M. Sugeno. A hierarchical subjective evaluation model using non-monotonic fuzzy measures and the Choquet integral. In M. Grabisch, T. Murofushi, and M. Sugeno, editors, *Fuzzy Measures and Integrals — Theory and Applications*, pages 375–391. Physica Verlag, 2000.
26. P. Miranda and M. Grabisch. Optimization issues for fuzzy measures. *Int. J. of Uncertainty, Fuzziness, and Knowledge-Based Systems*, 7(6):545–560, 1999.
27. T. Mori and T. Murofushi. An analysis of evaluation model using fuzzy measure and the Choquet integral. In *5th Fuzzy System Symposium*, pages 207–212, Kobe, Japan, 1989. In Japanese.
28. T. Murofushi and S. Soneda. Techniques for reading fuzzy measures (III): interaction index. In *9th Fuzzy System Symposium*, pages 693–696, Sapporo, Japan, May 1993. In Japanese.
29. D. Nettleton and V. Torra. A comparison of active set method and genetic algorithm approaches for learning weighting vectors in some aggregation operators. *Int. J. of Intelligent Systems*, 16(9):1069–1083, 2001.
30. T. Onisawa, M. Sugeno, Y. Nishiwaki, H. Kawai, and Y. Harima. Fuzzy measure analysis of public attitude towards the use of nuclear energy. *Fuzzy Sets & Systems*, 20:259–289, 1986.

31. E. Pap. *Null-Additive Set Functions.* Kluwer Academic, 1995.
32. D. Schmeidler. Subjective probability and expected utility without additivity. *Econometrica*, 57(3):571–587, 1989.
33. U. Schmidt. *Axiomatic Utility Theory under Risk.* Number 461 in Lectures Notes in Economics and Mathematical Systems. Springer Verlag, 1998.
34. L.S. Shapley. A value for n-person games. In H.W. Kuhn and A.W. Tucker, editors, *Contributions to the Theory of Games, Vol. II*, number 28 in Annals of Mathematics Studies, pages 307–317. Princeton University Press, 1953.
35. M. Sugeno. *Theory of fuzzy integrals and its applications.* PhD thesis, Tokyo Institute of Technology, 1974.
36. M. Sugeno and S.H. Kwon. A clusterwise regression-type model for subjective evaluation. *J. of Japan Society for Fuzzy Theory and Systems*, 7(2):291–310, 1995.
37. M. Sugeno and S.H. Kwon. A new approach to time series modeling with fuzzy measures and the Choquet integral. In *Int. Joint Conf. of the 4th IEEE Int. Conf. on Fuzzy Systems and the 2nd Int. Fuzzy Engineering Symp.*, pages 799–804, Yokohama, Japan, March 1995.
38. H. Tahani and J.M. Keller. Information fusion in computer vision using the fuzzy integral. *IEEE Tr. on Systems, Man, and Cybernetics*, 20(3):733–741, 1990.
39. Z. Wang and G.J. Klir. *Fuzzy measure theory.* Plenum, 1992.
40. Z. Wang, K.S. Leung, and J. Wang. A genetic algorithm for determining nonadditive set functions in information fusion. *Fuzzy Sets and Systems*, 102:462–469, 1999.
41. Z. Wang, K.S. Leung, M.L. Wong, J. Fang, and K. Xu. Nonlinear nonnegative multiregressions based on Choquet integrals. *Int. J.of Approximate Reasoning*, 25:71–87, 2000.
42. K. Xu, Z. Wang, and K.S. Leung. Using a new type of nonlinear integral for multiregression: an application of evolutionary algorithms in data mining. In *Proc. of IEEE Int. Conf. on Systems, Man and Cybernetics*, pages 2326–2331, 1998.

An Algorithm Based on Alternative Projections for a Fuzzy Measure Identification Problem

Hideyuki Imai[1], Daiki Asano[1], and Yoshiharu Sato[1]

Division of Systems and Information Engineering, Graduate School of Engineering, Hokkaido University, Nishi 8, Kita-ku Kita 13, Sapporo 060-8628, Japan

Abstract. Recently, there have been many applications of fuzzy measures to subjective evaluation models. In practical use, it is difficult to identify a fuzzy measure because many coefficients must be determined. Thus, there have been many efforts to develop an efficient algorithm to identify a fuzzy measure from examples. The properties of the solution set of a fuzzy measure identification problem have also been studied in detail.

In this chapter, we review the properties of the solution set and propose a new algorithm based on alternative convex projections.

1 Introduction

Fuzzy measures defined on a set of criteria or attributes represent their relative importance as well as their interaction. Thus, there are many applications of fuzzy measures to subjective evaluation models. For example, they are applied to analysis of public attitude towards the use of nuclear energy [10], subjective evaluation of printed color images [15], and subjective evaluation of an urban environment [5,9].

To define a fuzzy measure, we must determine coefficients for all subsets of the set of all criteria. In practical use, the number of criteria, which is denoted by M, is large. Thus, it is often difficult to define a fuzzy measure because $2^M - 1$ coefficients must be identified. Therefore, many researchers have tried to develop an efficient algorithm for determination of the coefficients [3,7,8,13,14]. In [6,8], it was proved that if we consider the quadratic error criterion, the problem is reduced to one of solving a convex quadratic programming problem, and it was shown that a complementary pivot algorithm can be applied to obtain a solution of a problem of identifying a fuzzy measure. In addition, an iterative algorithm has been proposed [3]. Moreover, some experimental results were reported to compare these methods [3,7].

These results make sense when the solution of the problem is unique. However, to obtain the conditions for uniqueness of a solution are not easy [6]. Thus, we attempted to specify the solution set of a fuzzy measure identification problem. At first, we review some properties of a solution set deduced from a convex programming problem, though they are well known. Next, we show the algorithm proposed by Boyle and Dykstra [1], which is based on alternative convex projections. However, in many cases, this algorithm can not be directly applied to a fuzzy measure identification problem. Thus, we modified the algorithm to a general situation, and we propose a new algorithm based on the modification,

2 The Choquet integral

Let $X = \{x_1, \ldots, x_M\}$ be the set of criteria or attributes, and let $\mathcal{F}$ be the family of all subsets of X. A monotonic fuzzy measure defined on $(X, \mathcal{F})$ is a real-valued function $\mu : \mathcal{F} \to [0, \infty)$ satisfying

$$\mu(\emptyset) = 0, \tag{1a}$$

$$A \subset B \text{ implies } \mu(A) \leq \mu(B) \text{ for } A, B \in \mathcal{F}. \tag{1b}$$

A fuzzy measure defined on $(X, \mathcal{F})$ consists of $2^M - 1$ real values satisfying the above conditions. As an alternative definition of a fuzzy measure, the condition

$$\mu(X) = 1 \tag{2}$$

is often added to the above conditions, and a fuzzy measure satisfying these three conditions is called a normalized fuzzy measure. In what follows, the value $\mu(A)$ for $A \in \mathcal{F}$ is called the measure of A.

The Choquet integral of a non-negative function $f : X \to [0, \infty)$ with respect to a fuzzy measure μ is defined by

$$(C)\int f d\mu = \int_0^\infty \mu(\{x \mid f(x) \geq r\}) dr.$$

Since we assume that X is a finite set, the Choquet integral becomes

$$(C)\int f d\mu = \sum_{m=2}^{M} \{f(x_{l_m}) - f(x_{l_{m-1}})\}\mu(A_m) + f(x_{l_1})\mu(A_1), \tag{3}$$

where $\{l_1, \ldots, l_M\}$ is the permutation of $\{1, \ldots, M\}$ so that

$$0 \leq f(x_{l_1}) \leq \cdots \leq f(x_{l_M}),$$

and

$$A_m = \{x_{l_m}, x_{l_{m+1}}, \ldots, x_{l_M}\}.$$

If we agree that $l_0 = 0$ and $f(x_0) = 0$ for any non-negative function f, equation (3) can be rewritten simply as

$$(C)\int f d\mu = \sum_{m=1}^{M} \{f(x_{l_m}) - f(x_{l_{m-1}})\}\mu(A_m). \tag{4}$$

An important advantage of the Choquet integral, with respect to other fuzzy integrals, is that it coincides with the Lebesgue integral when the measure is additive.

3 Fuzzy measure identification problem with the Choquet integral model

Let $y_1, \ldots, y_N$ be total evaluations of N objects (or by N individuals), and let $f_1(x_j), \ldots, f_N(x_j), j = 1, \ldots, M$, be their evaluations of attribute x_j. Our aim is to determine measures of all elements in $\mathcal{F}$ such that they minimize

$$||\boldsymbol{y} - \hat{\boldsymbol{y}}||^2, \tag{5}$$

where the norm $||\cdot||$ denotes the norm in N-dimensional space,

$$\begin{aligned} \boldsymbol{y} &= [y_1, \ldots, y_N]', \\ \hat{\boldsymbol{y}} &= [\hat{y}_1, \ldots, \hat{y}_N]', \\ \hat{y}_i &= (C)\int f_i d\mu, \end{aligned}$$

and $\boldsymbol{y}'$ denotes the transposed vector of a vector $\boldsymbol{y}$. When the Euclidean distance is used as $||\cdot||$, that is,

$$||\boldsymbol{y} - \hat{\boldsymbol{y}}||^2 = \sum_{i=1}^{N} (y_i - \hat{y}_i)^2, \tag{6}$$

it is shown that a fuzzy measure identification problem is reduced to one of solving the convex quadratic programming problem [8]. Thus, methods developed to solve a convex quadratic programming problem such as a complimentary pivot algorithm can be applied to determine a fuzzy measure [4,8]. Now we introduce some useful notations for a fuzzy measure identification problem [8]. Let μ_k be the measure of the set

$$\{x_l \in X \mid \delta_l^k = 1, l = 1, \ldots, M\},$$

where $\delta_M^k \delta_{M-1}^k \cdots \delta_1^k$ is the dyadic system representation of an integer k, that is,

$$k = 2^{M-1}\delta_M^k + 2^{M-2}\delta_{M-1}^k + \cdots + 2\delta_2^k + \delta_1^k, \delta_l^k \in \{0, 1\}. \tag{7}$$

By the notation,

$$\begin{aligned} \mu_1 &= \mu(\{x_1\}), \\ \mu_2 &= \mu(\{x_2\}), \\ \mu_3 &= \mu(\{x_1, x_2\}), \\ &\vdots \\ \mu_{2^M-1} &= \mu(X). \end{aligned}$$

Moreover, let $A = (a_{ij})$ be an $N \times (2^M - 1)$ matrix where

$$a_{ij} = \begin{cases} f_i(x_{l_k^i}) - f_i(x_{l_{k-1}^i}), & j = \sum_{s=k}^{M} 2^{l_s^i - 1}, k = 1, \ldots, M, \\ 0, & \text{otherwise}, \end{cases}$$

and $\{l_1^i, \dots, l_M^i\}$ is the permutation of $\{1, \dots, M\}$ so that

$$0 \le f_i(x_{l_1^i}) \le \cdots \le f_i(x_{l_M^i}).$$

By these notations, equation (5) becomes

$$(\boldsymbol{y} - A\boldsymbol{\mu})'(\boldsymbol{y} - A\boldsymbol{\mu}), \tag{8}$$

where $\boldsymbol{\mu} = [\mu_1, \dots, \mu_{2^M-1}]'$. Thus, the problem of identifying a fuzzy measure is to obtain a vector $\boldsymbol{\mu}$ minimizing (8) subject to

$$\mu_k \ge 0, \qquad k = 1, \dots, 2^M - 1, \tag{9a}$$

$$\mu_k > \mu_{k'}, \text{ if } \delta_l^k \ge \delta_l^{k'}, \forall l = 1, \dots, M, \tag{9b}$$

where $\delta_M^k \cdots \delta_1^k$ is the dyadic representation of an integer k as in (7). Inequality (9b) corresponds to condition (1b). It should be noted that the number of constraints is $M(2^{M-1} - 1)$. Let Ω be the set of vectors whose components satisfy these inequalities, and a vector $\boldsymbol{\mu} \in \Omega$ minimizing (8) is called a solution of a fuzzy measure identification problem.

4 The Set of solutions of a fuzzy measure identification problem

In this section, we present some properties of the set of solutions of a fuzzy measure identification problem.

Proposition 1 *Let Ω be the set of vectors whose components satisfy conditions (9a) and (9b). Thus, there exists a unique element*

$$\hat{\boldsymbol{y}} \in A\Omega = \{A\boldsymbol{\mu} \in \boldsymbol{R}^N \mid \boldsymbol{\mu} \in \Omega\},$$

such that

$$\inf_{\boldsymbol{z} \in A\Omega} (\boldsymbol{y} - \boldsymbol{z})'(\boldsymbol{y} - \boldsymbol{z}) = (\boldsymbol{y} - \hat{\boldsymbol{y}})'(\boldsymbol{y} - \hat{\boldsymbol{y}}),$$

because $A\Omega$ is a convex set. The element $\hat{\boldsymbol{y}}$ is called the convex projection of $\boldsymbol{y}$ onto $A\Omega$.

Therefore, there exists at least one element belonging to Ω that minimizes (8). Thus, the set of solutions, which is denoted by Θ, is not empty.

Proposition 2 *The set of solutions Θ is a polyhedral convex set in $\boldsymbol{R}^{2^M-1}$, where a polyhedral convex set is a set which can be expressed as the intersection of some finite collection of closed half planes.*

Proofs of these Propositions are found in Rockafellar [12] or Youla [16].

Proposition 3 *The set of solutions Θ is a polytope if and only if it holds that*

$$\max_{1\le i\le N}(\min_{1\le j\le M} f_i(x_j)) > 0, \tag{10}$$

where a polytope is the convex hull of finitely many points.

Proof. Condition (10) means that there exists $i_1 \in \{1, \dots, N\}$ such that

$$\min_{1\le j\le M} f_{i_1}(x_j) > 0.$$

Thus, by the definition of the Choquet integral, it holds that

$$f_{i_1}(x_j)\mu_{2^M-1} = f_{i_1}(x_j)\mu(X) \le \hat{y}_{i_1}.$$

Therefore, by monotonicity of a fuzzy measure, all components of $\boldsymbol{\mu} \in \Theta$ are bounded. Thus, we have shown that Θ is a polytope .

Now, suppose that condition (10) does not hold. Since f is a non-negative function, it yields that

$$\min_{1\le j\le M} f_i(x_j) = 0, i = 1, \dots, N.$$

Therefore, it is clear that $\mu_{2^M-1} = \mu(X)$ is not bounded. □

In a normalized fuzzy measure identification problem, the set of solutions is always a polytope because the all solutions are bounded. In regression analysis, the unknown regression coefficient is uniquely determined if the rank of the design matrix is less than or equal to the sample size. A similar result holds in a fuzzy measure identification problem as follows.

Proposition 4 *If rank$(A) = 2^M - 1$, the solution is unique, and $\tilde{\boldsymbol{\mu}} = (A'A)^{-1}A'\hat{\boldsymbol{y}}$*

Miranda and Grabisch [6] investigated the condition that a fuzzy measure identification problem has a unique solution, and show an example that the condition in the proposition is not a sufficient one.

5 An algorithm based on alternative projections

In this section, we propose a new algorithm based on alternative convex projections when the quadratic error criterion (6) is used.

To obtain the solution of a fuzzy measure identification problem, some algorithms based on a convex quadratic programming problem have been developed [4,8]. We propose another method for searching for optimum solutions of a fuzzy measure identification problem. This method is based on alternative convex projections proposed by Boyle and Dykstra [1]. They showed that if the constraint region can be expressed as a finite intersection of simple convex regions, one can get the projection onto the intersection obtained by performing a series of convex projections onto simpler regions.

At first, we show the least squares estimate in a linear model,

$$y = A\boldsymbol{\mu} + \epsilon, \tag{11}$$

where $\boldsymbol{\mu}$ is constrained to satisfy the restriction $\boldsymbol{\mu} \in \Omega$, and the matrix $A'A$ is full rank.

Let $\hat{\boldsymbol{\mu}} \in \boldsymbol{R}^{2^M-1}$ be the least squares estimate without restrictions in the linear model (11). Thus, the least squares estimate $\tilde{\boldsymbol{\mu}} \in \boldsymbol{R}^{2^M-1}$ under the restriction $\boldsymbol{\mu} \in \Omega$ is the projection $\hat{\boldsymbol{\mu}}$ onto Ω, where the metric is determined by the inner product

$$\langle \boldsymbol{\mu}_1, \boldsymbol{\mu}_2 \rangle = \boldsymbol{\mu}_1' A' A \boldsymbol{\mu}_2, \boldsymbol{\mu}_1, \boldsymbol{\mu}_2 \in \boldsymbol{R}^{2^M-1}.$$

Therefore, the aim is to obtain $\tilde{\boldsymbol{\mu}} \in \boldsymbol{R}^{2^M-1}$ such that

$$\tilde{\boldsymbol{\mu}} = \arg\min_{\boldsymbol{\mu} \in \Omega} (\boldsymbol{\mu} - \hat{\boldsymbol{\mu}})' A' A (\boldsymbol{\mu} - \hat{\boldsymbol{\mu}}).$$

If $\Omega = \Omega_1 \cap \cdots \cap \Omega_r$, where each Ω_i is a convex set, Boyle and Dykstra [1] proposed an algorithm for the solution of the problem. The algorithm consists of 2 cycles. For convenience, we set $\boldsymbol{\mu}_{01} = \hat{\boldsymbol{\mu}}$.

Cycle 1 :

1. Project $\boldsymbol{\mu}_{0r} (= \hat{\boldsymbol{\mu}})$ onto Ω_1 and obtain

$$\boldsymbol{\mu}_{11} = \hat{\boldsymbol{\mu}} + I_{11}.$$

2. Project $\boldsymbol{\mu}_{01}$ onto Ω_2 and obtain

$$\boldsymbol{\mu}_{12} = \boldsymbol{\mu}_{11} + I_{12} = \hat{\boldsymbol{\mu}} + I_{11} + I_{12}.$$

$$\vdots$$

r. Project $\boldsymbol{\mu}_{1,r-1}$ onto Ω_r and obtain

$$\boldsymbol{\mu}_{1r} = \boldsymbol{\mu}_{1,r-1} + I_{1r} = \hat{\boldsymbol{\mu}} + I_{11} + I_{12} + \cdots + I_{1r}.$$

In Cycle 1, I_{1i} means the difference between $\boldsymbol{\mu}_{1i}$ and its convex projection. After the first cycle, the second cycle proceeds as follows.

Cycle 2 :

1. Project $\boldsymbol{\mu}_{1r} - I_{11}$ onto Ω_1 to obtain

$$\boldsymbol{\mu}_{21} = \boldsymbol{\mu}_{1r} - I_{11} + I_{21} = \hat{\boldsymbol{\mu}} + I_{21} + I_{12} + \cdots + I_{1r}.$$

2. Project $\boldsymbol{\mu}_{21} - I_{12}$ onto Ω_2 to obtain

$$\boldsymbol{\mu}_{22} = \boldsymbol{\mu}_{21} - I_{12} + I_{22} = \hat{\boldsymbol{\mu}} + I_{21} + I_{22} + I_{13} + \cdots + I_{1r}.$$

$$\vdots$$

r. Project $\boldsymbol{\mu}_{2,r-1} - I_{1r}$ onto Ω_r to obtain

$$\boldsymbol{\mu}_{2r} = \boldsymbol{\mu}_{2,r-1} - I_{1r} + I_{2r} = \hat{\boldsymbol{\mu}} + I_{21} + +I_{22} + \cdots + I_{2r}.$$

Continuing Cycle 2 generates the infinite sequences $\boldsymbol{\mu}_{ni}$ and I_{ni}, where $n \geq 1$ and $1 \leq i \leq r$. We see that the following relations hold for $n \geq 1$ and $i = 2, \ldots, r$.

$$\begin{aligned} \boldsymbol{\mu}_{n-1,r} - \boldsymbol{\mu}_{n1} &= I_{n-1,1} - I_{n1}, \\ \boldsymbol{\mu}_{n,i-1} - \boldsymbol{\mu}_{ni} &= I_{n-1,i} - I_{ni}, \end{aligned}$$

where for convenience we set $\boldsymbol{\mu}_{r1} = \hat{\boldsymbol{\mu}}$ and $I_{0i} = 0$ for all i. Thus, in Cycle 2, I_{ni} means the difference between $\boldsymbol{\mu}_{n,i-1} - I_{n-1,i}$ and its convex projection.

For the algorithm, the following was shown. The proof of this proposition is shown in [1].

Proposition 5 (Boyle and Dykstra [1]) *For any $1 \leq i \leq r$, the sequence $\{\boldsymbol{\mu}_{ni}\}$ converges to $\tilde{\boldsymbol{\mu}}$, that is,*

$$(\boldsymbol{\mu}_{ni} - \tilde{\boldsymbol{\mu}})'(\boldsymbol{\mu}_{ni} - \tilde{\boldsymbol{\mu}}) \to 0 \ \textit{as} \ n \to \infty.$$

In a fuzzy measure identification problem, the matrix $A'A$ is not always nonsingular. Thus, the above algorithm can not be applied directly since $\langle \boldsymbol{\mu}_1, \boldsymbol{\mu}_2 \rangle = \boldsymbol{\mu}_1' A' A \boldsymbol{\mu}_2$ is not an inner product when $A'A$ is singular.

However, if we consider the subspace $\{\boldsymbol{\mu} \in \boldsymbol{R}^{2^M-1} \mid \boldsymbol{\mu} \in \mathrm{ran}(A')\}$, where $\mathrm{ran}(A')$ denotes the range of A', $\langle \boldsymbol{\mu}_1, \boldsymbol{\mu}_2 \rangle = \boldsymbol{\mu}_1' A' A \boldsymbol{\mu}_2$ is the inner product in $\mathrm{ran}(A')$. Thus, we can show the following.

Proposition 6 *Let $\hat{\boldsymbol{\mu}} \in ran(A')$ be the least squares estimate without restrictions in the linear model (11). Thus, the least squares estimate $\tilde{\boldsymbol{\mu}} \in ran(A')$ under the restriction $\boldsymbol{\mu} \in \Omega \cap ran(A')$ is the projection $\hat{\boldsymbol{\mu}}$ onto $\Omega \cap ran(A')$, where the metric is determined by the inner product*

$$\langle \boldsymbol{\mu}_1, \boldsymbol{\mu}_2 \rangle = \boldsymbol{\mu}_1' A' A \boldsymbol{\mu}_2, \boldsymbol{\mu}_1, \boldsymbol{\mu}_2 \in ran(A'). \tag{12}$$

The set of all least squares estimates with restriction in $\boldsymbol{R}^{2^M-1}$ is all solutions of

$$A\boldsymbol{\mu} = A\tilde{\boldsymbol{\mu}}, \text{ subject to } \boldsymbol{\mu} \in \Omega$$

Therefore, if the convex projection $\tilde{\boldsymbol{\mu}} \in \boldsymbol{R}^{2^M-1}$ is obtained, the set of all least squares estimates under the constraint is the feasible region of the linear programming problem.

The least squares estimate of $\boldsymbol{\mu}$ without restriction is

$$\hat{\boldsymbol{\mu}} = (A'A)^{\dagger} A'\boldsymbol{y},$$

where $A^{\dagger}$ denotes the Moore-Penrose inverse matrix of matrix A. Thus, by Proposition 6, the problem is to obtain the convex projection of $\hat{\boldsymbol{\mu}}$ onto $\Omega \cap \mathrm{ran}(A')$, where the metric on $\mathrm{ran}(A')$ is determined by the inner product (12). To do this, let C be an $M(2^{M-1} - 1) \times (2^M - 1)$ matrix representing constraints (9a) and (9b), that is,

$$\Omega = \{\boldsymbol{\mu} \in \mathrm{ran}(A') \subset \boldsymbol{R}^{2^M - 1} \mid C\boldsymbol{\mu} \geq \boldsymbol{b}\}.$$

Unless we consider a normalized fuzzy measure, the vector $\boldsymbol{b}$ is the zero vector.

The restriction

$$C\boldsymbol{\mu} \geq \boldsymbol{b},$$

is expressed as an intersection of convex regions,

$$\boldsymbol{c}_i'\boldsymbol{\mu} \geq b_i, i = 1, \ldots, r,$$

where $r = M(2^{M-1} - 1)$, $\boldsymbol{b} = [b_1, \ldots, b_r]'$ and

$$C = \begin{bmatrix} \boldsymbol{c}_1' \\ \vdots \\ \boldsymbol{c}_r' \end{bmatrix}.$$

Thus, for the algorithm, the projection onto

$$\Omega_i = \{\boldsymbol{\mu} \in \mathrm{ran}(A') | \boldsymbol{c}_i'\boldsymbol{\mu} \geq b_i\},$$

is needed where the metric is determined by the inner product (12).

Let P_{Ω_i} be the convex projector onto Ω_i. Thus,

$$P_{\Omega_i}\boldsymbol{\mu} = \begin{cases} \boldsymbol{\mu}, & \boldsymbol{c}_i'\boldsymbol{\mu} \geq b_i, \\ \boldsymbol{\mu}_*, & \boldsymbol{c}_i'\boldsymbol{\mu} < b_i, \end{cases}$$

where $\langle \boldsymbol{\mu}_* - \boldsymbol{\mu}, \boldsymbol{\mu}_* - \boldsymbol{\mu} \rangle = \inf_{\boldsymbol{\mu}_1 \in \mathrm{ran}(A'), \boldsymbol{c}_i'\boldsymbol{\mu}_1 = b_i} \langle \boldsymbol{\mu}_1 - \boldsymbol{\mu}, \boldsymbol{\mu}_1 - \boldsymbol{\mu} \rangle$.

Proposition 7 *Let t be the rank of matrix A, and*

$$A'A = P \begin{bmatrix} \Lambda & O \\ O & O \end{bmatrix} P',$$

where Λ is $t \times t$ diagonal matrix whose diagonal elements are the singular values of matrix $A'A$, P is an orthogonal matrix, and O denotes the zero matrix. Then,

$$\mu_* = A' \begin{bmatrix} \Lambda & O \\ O & I \end{bmatrix} P' \begin{bmatrix} \eta_{1*} \\ \eta_{2*} \end{bmatrix} + \boldsymbol{\mu}_1 \tag{13}$$

minimizes

$$(\boldsymbol{\mu} - \boldsymbol{\mu}_1)' A' A (\boldsymbol{\mu} - \boldsymbol{\mu}_1), \boldsymbol{\mu}_1 \in ran(A')$$

subject to

$$\boldsymbol{c}' \boldsymbol{\mu} = b_i, \quad \boldsymbol{\mu} \in ran(A'),$$

where

$$\boldsymbol{\eta}_{1*} = d_1 (b_i - \boldsymbol{c}' \boldsymbol{\mu}_1) \boldsymbol{\omega}_1, \tag{14a}$$

$$\boldsymbol{\eta}_{2*} = (b_i - \boldsymbol{c}' \boldsymbol{\mu}_1) \boldsymbol{d}_2, \tag{14b}$$

$$\begin{bmatrix} \boldsymbol{\omega}_1 \\ \boldsymbol{\omega}_2 \end{bmatrix} = \begin{bmatrix} \Lambda^{-1} & O \\ O & I \end{bmatrix} P' A \boldsymbol{c}_i,$$

$$\begin{bmatrix} d_1 & \boldsymbol{d}_2' \\ \boldsymbol{d}_2 & D \end{bmatrix} = \begin{bmatrix} \boldsymbol{\omega}_1' \boldsymbol{\omega}_1 & \boldsymbol{\omega}_2' \\ \boldsymbol{\omega}_2 & O \end{bmatrix}^{\dagger}.$$

Proof. Since both $\boldsymbol{\mu}$ and $\boldsymbol{\mu}_1$ are in ran(A'), there exists $\boldsymbol{\xi} \in \boldsymbol{R}^N$ such that $\boldsymbol{\mu} - \boldsymbol{\mu}_1 = A' \boldsymbol{\xi}$. Thus, the problem is to minimize

$$\boldsymbol{\xi}' A' A A' A \boldsymbol{\xi}$$

subject to

$$(A \boldsymbol{c}_i) \boldsymbol{\xi} = \boldsymbol{c}' \boldsymbol{\mu}_1.$$

By putting $\boldsymbol{\eta} = \begin{bmatrix} \Lambda^{-1} & O \\ O & I \end{bmatrix} P' \boldsymbol{\xi}$, the criterion and the constraint are

$$\boldsymbol{\eta}' \begin{bmatrix} I_r & O \\ O & O \end{bmatrix} \boldsymbol{\eta} \ , \tag{15}$$

$$\left(\begin{bmatrix} \Lambda^{-1} & O \\ O & I \end{bmatrix} P' A \boldsymbol{c}_i, \right)' \boldsymbol{\xi} = b_i - \boldsymbol{c}_i' \boldsymbol{\mu}_1, \tag{16}$$

respectively. Applying the result found in Rao [11], $\boldsymbol{\eta}_* = \begin{bmatrix} \boldsymbol{\eta}_{1*} \\ \boldsymbol{\eta}_{2*} \end{bmatrix}$ defined in (14a) and (14b) minimizes (15) subject to (16).

It should be noted that if $b_i = 0$, then

$$\boldsymbol{\eta}_{1*} = d_1 (\boldsymbol{\omega}_1 \boldsymbol{c}_i') \boldsymbol{\mu}_1$$

$$\boldsymbol{\eta}_{2*} = (\boldsymbol{d}_1 \boldsymbol{c}_i') \boldsymbol{\mu}_1$$

and

$$\boldsymbol{\mu} = \left(A' \begin{bmatrix} \Lambda^{-1} & O \\ O & I \end{bmatrix} P \begin{bmatrix} d_1 (\boldsymbol{\omega}_1 \boldsymbol{c}_i') \\ d_2 \boldsymbol{c}_i' \end{bmatrix} + I \right) \boldsymbol{\mu}_1.$$

Thus, the algorithm based on convex projections $P_{\Omega_1}, \ldots, P_{\Omega_r}$ is as follows.

Cycle 1 :

$$\begin{array}{lll} 1. & \boldsymbol{\mu}_{11} = P_{\Omega_1}\hat{\boldsymbol{\mu}}, & I_{11} = \boldsymbol{\mu}_{11} - \hat{\boldsymbol{\mu}}. \\ 2. & \boldsymbol{\mu}_{12} = P_{\Omega_2}\boldsymbol{\mu}_{11}, & I_{12} = \boldsymbol{\mu}_{12} - \boldsymbol{\mu}_{11}. \\ & \vdots & \\ r. & \boldsymbol{\mu}_{1r} = P_{\Omega_r}\boldsymbol{\mu}_{1,r-1}, & I_{1r} = \boldsymbol{\mu}_{1r} - \boldsymbol{\mu}_{1,r-1}. \end{array}$$

Cycle 2 :

$$\begin{array}{lll} 1. & \boldsymbol{\mu}_{21} = P_{\Omega_1}(\boldsymbol{\mu}_{1r} - I_{11}), & I_{21} = \boldsymbol{\mu}_{21} - \boldsymbol{\mu}_{1r} + I_{11}. \\ 2. & \boldsymbol{\mu}_{22} = P_{\Omega_2}(\boldsymbol{\mu}_{21} - I_{12}), & I_{22} = \boldsymbol{\mu}_{22} - \boldsymbol{\mu}_{21} + I_{12}. \\ & \vdots & \\ r. & \boldsymbol{\mu}_{2r} = P_{\Omega_r}(\boldsymbol{\mu}_{2,r-1} - I_{1r}), & I_{2r} = \boldsymbol{\mu}_{2,r} - \boldsymbol{\mu}_{2,r-1} + I_{1r}. \end{array}$$

By Proposition 5, we see that the sequence $\{\boldsymbol{\mu}_{ni}\}, 1 \leq i \leq r$ converges to $\hat{\boldsymbol{\mu}} \in \mathrm{ran}(A')$. Once the convex projections onto $\Omega \cap \mathrm{ran}(A')$ is obtained, the set of all solutions of the problem is obtained by solving

$$A\boldsymbol{\mu} = A\hat{\boldsymbol{\mu}}, \text{ subject to } \boldsymbol{\mu} \in \Omega.$$

Because Boyle and Dykstra [1] shows that the algorithm is valid for a general Hilbert space rather than $\boldsymbol{R}^N$, we can use the proposed algorithm in more general situations [2].

6 Conclusions

In this chapter, we propose a convex projection-based algorithm for solving a fuzzy measure identification problem.

In the complementary pivot algorithm, it is guaranteed that iterations are always finite. On the other hand, the proposed algorithm generally requires infinite iterations. However, in many cases, it is reported that convergence speed to the optimal solution is high and the number of iterations is not so large[1].

In this chapter, we have proposed the algorithm and have not yet apply it to practical data. Thus, in future works, we will compare the results obtained by using proposed algorithm with those obtained by using the complementary pivot algorithm in practical data sets, especially with large sample size.

References

1. Boyle, J. P., Dykstra, R. (1985) A method of finding projections onto the intersection of convex sets in Hilbert spaces. In: Dykstra, R., Robertson, T., and Wright, F. T. (Eds.) Advances in Order Restricted Statistical Inference, Springer-Verlag, Berlin Heidelberg, 28–47.

2. Combettes, P. L. (1993) The foundation of set theoretic estimation. Proceedings of the IEEE, **81**, 182–208.
3. Grabisch, M. (1995) A new algorithm for identifying fuzzy measures and its application to pattern recognition. In:Proceedings of 4th FUZZ-IEEE, Yokohama, 145–151.
4. Grabisch, M., Nguyen, H. T., Walkerand, E. A.(1995) Fundamentals of Uncertainty Calculi with Application to Fuzzy Inference. Kluwer Academic.
5. Iwamoto, N., Ryouke, M., Nakamori, Y. (1997) A fuzzy model of subjective urban environment evaluation. Journal of Japan Society for Fuzzy Theory and Systems **9**, 279–286. (in Japanese).
6. Miranda, P., Grabisch, M. (1999) Optimization issues for fuzzy measures. International Journal of Uncertainty, Fuzziness and Knowledge-Based Systems **7**, 545–560.
7. Mori, T., Murofushi, T. (1989) An analysis of evaluation model using fuzzy measure and the Choquet integral. In: Proceedings of 5th Fuzzy System Symposium, Kobe, 207–212. (in Japanese).
8. Nakamori, Y. (1998) Identification and interpretation of fuzzy measures. Journal of Japan Society for Fuzzy Theory and Systems **10**, 215–224. (in Japanese).
9. Nakamori, Y., Iwamoto, N., Naito, M. (1995) Analysis of subjective urban environmental evaluation structure by the Choquet integral model. Environmental Science **8**, 11–24. (in Japanese).
10. Onisawa, T., Sugeno, M., Nishiwaki, Y., Kawai, H., Harima, Y. (1986) Fuzzy measure analysis of public attitude towards the use of nuclear energy. Fuzzy Sets and Systems **20**, 259–289.
11. Rao, C. R. (1973) Linear Statistical Inference and Its Application. Wiley, New York.
12. Rockafellar, R. T. (1970) Convex Analysis. Princeton University Press, Princeton.
13. Takahagi, E., Murofushi, T. (1995) On identification of a fuzzy measure. In: Proceedings of 5th Intelligent System Symposium, Tokyo, 463–468. (in Japanese).
14. Tanaka, A., Murofushi, T. (1989) A learning model using fuzzy measure and the Choquet integral. In: Proceedings of 5th Fuzzy System Symposium, Kobe, 213–217. (in Japanese).
15. Tanaka, K., Sugeno, M. (1991) A study on subjective evaluations of printed color images. Journal of Approximate Reasoning **5**, 213–222.
16. Youla, C. K. (1987) Mathematical theory of image restoration by the method of convex projections. In: Stark, H. (Eds.) Image Recovery: Theory and Application, Academic Press, San Diego, 29–77.

Combining Information Fusion with String Pattern Analysis: A New Method for Predicting Future Purchase Behavior

Yukinobu Hamuro[1], Naoki Katoh[2], Edward H. Ip[3], Stephane L. Cheung[3], and Katsutoshi Yada[4]

[1] Faculty of Business Administration, Osaka Sangyo University, Daito, Osaka, Japan
[2] Department of Architecture and Architectural Systems, Kyoto University Yoshida-Honmachi, Sakyo-ku, Kyoto, 606-8501 Japan
[3] Information and Operations Management, Marshall School of Business, University of Southern California, Los Angeles, CA, USA
[4] Faculty of Commerce, Kansai University, Yamatemachi, Suita, Osaka, 564-8680 Japan

Abstract. We develop a new method for extracting useful knowledge from individual purchase history of customers by combining information fusion techniques with a data mining tool – string pattern analysis. We demonstrate through several case studies how the method helps firms predict how and when a customer is likely to switch from one brand to another, who is most likely to become loyal to a brand, and when a customer is likely to defect to a competitor. The method comprises several phases of data fusion processes. First, as preprocessing, it transforms purchase history of individual customers into a string of symbols, each of which represents a purchased brand, and then it maps a set of symbols into a much smaller set of new symbols in order to enhance the predictability of the method. Second, as the method produces simple and effective rules described by regular pattern over the set of symbols that effectively distinguish given positive and negative sets of strings. The string pattern analysis retains temporal information in a purchase sequence, an advantage that is not usually enjoyed by other data analytical methods. Finally, we devise a decision tree algorithm and a weighted majority voting algorithm based two-dimensional region rules to aggregate predictors and generate interpretable rules. This is also a data fusion process. The method is illustrated by several applications: brand switching management, customer attrition management, and loyalty program management. The successful applications demonstrate the effective combination of information fusion and string pattern analyses in addressing real business problems.

1 Introduction

The rapid development of modern information technologies has led to important progress in the collecting, processing, and dissemination of data. This has resulted in the accumulation of tremendous amount of business data in corporate databases. The ability to leverage the data has become a key success factor in an increasingly competitive global market. The availability of detailed customer data and advances in technology for warehousing and mining data (e.g. [20,21]) enable companies to better understand and service their customers. Knowledge discovery in databases or data mining is a new technology or methodology that seeks to automatically extract meaningful knowledge from business data [2,18,23]. However, as explained in the introductory chapter of this volume, it is not a trivial task to extract meaningful business knowledge from such raw data. From our experiences, one of the keys to the success in this task is in transforming the raw data into appropriately summarized information. This process itself can be viewed as information fusion [1,22]. The purpose of this chapter is to demonstrate how useful knowledge can be extracted through a combination of information fusion and data mining techniques. Specifically, we developed methods for string pattern analyses for predicting future customer behavior based on customer purchase history.The work in this chapter is built upon several recent investigations in data mining techniques for large transaction database [8,9,13,4].

The proposed method is motivated by new techniques that were developed for the analysis of string patterns, which are commonly found in areas such as genetic sequencing. In this article, we emphasize string pattern methods for consumer purchase data. To discover rules about purchase behavior, it is often important to analyze not just cross-sectional data but purchase history. In order to directly make use of time-series purchase history data, we transform purchase history into a string of symbols. Each symbol may represent a purchased brand, or a categorized profitability value. The analysis of strings makes use of information that is contained in the time series and thus has an advantage over methods that only use static numerical and/or categorical information such as profitability per visit and number of visits. As demonstrated in this article, the proposed method discovers knowledge which may otherwise not be uncovered by traditional analytical methods.

We extend a machine discovery system BONSAI that was originally developed in the field of molecular biology by Shimozono et al. [3,19]. Using a technique called *alphabet indexing*, BONSAI produces a decision tree over regular patterns[1] from a given positive and a given negative set of strings. The core of the system is an engine that generates a decision tree which accurately classifies positive and negative examples. The alphabet indexing system transforms original strings via a mapping from an alphabet with the large number of symbols to another alphabet with fewer symbols. The transformation does not lose any positive or negative information of the given

[1] The term "regular pattern" is used in this chapter following the definition introduced in [3,19]. A regular pattern π is a string of the form $\pi = w_0 x_1 w_1 x_1 w_2 x_2 \cdots x_n w_n$, where each w_i is a constant string and each x_i is a variable that matches any string. Variable x_i is often denoted simply by $*$. Hence the above pattern defines any string containing substrings $w_0, w_1, \ldots, w_n$ in that order.

examples. It was observed in [19] that, in a biomedical application, the use of an appropriate alphabet indexing scheme can increase accuracy and simplifies hypotheses. Furthermore, [19] reported that an alphabet indexing scheme, called transmembrane identification, matches domain knowledge, and can be interpreted in terms of a hydropathy index.

There are limitations to the original version of BONSAI. For example, regular patterns attached to the nodes of decision trees are limited to substrings, which are of the form $x\alpha y$ where x and y are variables and α is a substring taken from positive and negative examples. This limitation was recently addressed in [10]. We develop an extended version of BONSAI which has the following new features:

1) While original BONSAI generates a decision tree over regular patterns which are limited to substrings, we extend it to subsequences based on the work of [10].
2) We generate rules which contain not only regular patterns but numerical attributes such as age, number of visits, profit and so on, while BONSAI uses only string patterns as predictive attributes. The latter feature allows us to incorporate various types of attributes into the learning process.
3) Because it is usually believed that the most recently purchased brand is closely related to the next purchase, we extend regular pattern to take into account the position where a certain pattern occurs within the string.
4) Hoping to generate more interpretable rules and to give higher accuracy than original decision tree based BONSAI, we implement majority voting based on two-dimensional region rules as an alternative which we call region-based BONSAI or simply region-BONSAI. In fact, as will be shown in Section 3, region-BONSAI enhances the prediction accuracy and sometimes generates more interpretable rules.

According to Torra [22], three information fusion processes can be distinguished: 1) preprocessing of data, 2) model building and 3) information extraction. In the knowledge discovery procedure of our method, encoding of the original purchase history into a string of symbols, the aggregation of predictors using majority voting in region-BONSAI, and the alphabet indexing that reduces the model size, and also enhances the prediction ability can be viewed as the information fusion process.

In our method, temporal information is re-identified and encoded into strings of symbols, i.e., it combines purchase history of an individual customer recorded in multiple transactions in a database into one string of symbols each of which represents an object such as a brand purchased, a degree of frequency of visit, etc. This is exactly the preprocessing of the data. In region-BONSAI based on majority voting, we choose a subset of decision rules to predict whether the target attribute of a given record is positive or negative by a weighted majority decision, with weights appropriately assigned to each selected rule, hoping that the resulting decision is better (in terms of accuracy) than would be possible if any of decision rules is individually used. Thus, such decision mechanism can be viewed as aggregation operators to fuse data models (see [22]), and thus contributes to the fusion precess in model building. Alphabet indexing used in our method can be viewed as a preprocessing of data which aggregates a set of symbols into a smaller set of symbols. It is reported that the procedure enhances accuracy ([3,8]). In the original method in [3,19] as well as ours, the best alphabet indexing is sought during the process of finding the best prediction model. In this sense, the alphabet indexing contributes to the fusion precess in model building as well. The alphabet indexing in the prediction model obtained sometimes produces useful knowledge as reported in [3,19]. In this sense, it also contributes to the fusion process in information extraction to some extent.

We applied string pattern analysis to three data sets of customer purchase history that were obtained from a drugstore chain in Japan. The experiments show that the proposed method produces meaningful and interpretable rules concerning customer purchase behavior. While we do extend the machine learning system BONSAI by adding new features to it, the main theme of this chapter is in demonstrating the effectiveness of combining information fusion and string pattern techniques for the analysis of future purchase behavior.

2 String Pattern Analysis

In this section we describe string pattern analysis, the method developed for analyzing large purchase history database. The method first applies information fusion to represent a sequence of purchase history as character strings and then extract rules from the string data. We describe the information fusion process, and then the algorithm for extracting rules.

2.1 Fusion from sales history data to string pattern

Let us start with a simple example. Figure 1(a) represents the purchase history data collected on an individual customer. The table shows which brand the customer purchased and on what date. The brand is represented as a code like "10550". Purchase history data may contain other important information such as price, quantity, discount at purchase, and so on.

In transforming the history data, first we construct a mapping between brand code and the set of alphabets (Figure 1(b)). Then we concatenate the sequence of brands, now in the form of a string of alphabets, in the same order as the sequence of purchase. For example, a customer whose string pattern 'AABCB' implies that he bought brand A two times, then switched to brand B, C, and back to B again (Figure 1(c)). The information fusion process retains valuable time related information that might otherwise have been lost because of data summarization. For example, purchase history can be summarized into variables such as the number of brands a customer purchased in a month. For secondary users of the transformed data, it is difficult to recover detailed attributes from the summarized sales history data. In the above example, secondary users will not be able to retrieve information such as whether the switching between brands is regular or random. The fusion process described above does lose *interval* information, i.e., we do not know how many days elapse between consecutive purchases. We shall deal with this problem by introducing a new attribute concerning interval information and then by creating a symbol string representing a sequence of intervals between two consecutive visits (see Section 3.3 for the details).

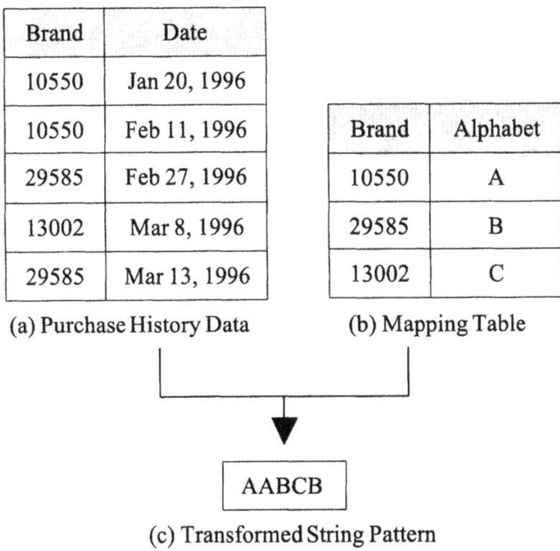

Brand	Date
10550	Jan 20, 1996
10550	Feb 11, 1996
29585	Feb 27, 1996
13002	Mar 8, 1996
29585	Mar 13, 1996

(a) Purchase History Data

Brand	Alphabet
10550	A
29585	B
13002	C

(b) Mapping Table

(c) Transformed String Pattern

Fig. 1. Transformation from purchase history to string pattern

We develop ways for transforming different types of history data to string pattern. These methods are outlined below.

a) Categorical attributes: Let D denote the purchase history data of an individual customer. Suppose the set D contains a set of attributes A_i, (denoted by $\mathcal{A}$), and A_t, where A_t is a time related quantity such as date. For each $A_i \in \mathcal{A}$, create a one-to-one mapping $\Sigma_i \rightarrow \Gamma_i$, where Σ_i is a domain of attribute A_i, and Γ_i is the set of alphabets. The transformation replaces each value of A_i in D with the corresponding alphabet using mapping table $\Sigma_i \rightarrow \Gamma_i$. Concatenate all mapped character in the order of A_t on each A_i. Accordingly, each customer is given multiple string patterns S_i for each attribute A_i.

b) Numerical attributes: Numerical attributes often occur in transaction – price, number of visits, quantity of items purchased are all examples of numerical attributes. To transform numerical attributes, discretize the range of values taken by the attribute. Then create the mapping table $N_i \rightarrow \Gamma_i$, where N_i is a range of the numerical value. One possible way to discrete the range is to use "buckets" of values such that each bucket has approximately the same number of cases. Other methods appropriate for the situation at hand can also be considered.

c) Combinations: It is also possible to combine strings of categorical and numerical attributes. Modify the mapping table to $(\Sigma, N) \rightarrow \Gamma$. Figure 2 shows a sample of such mapping tables, in which an alphabet has two different meanings - alphabets correspond to brands purchased, and lower and upper case of alphabets respectively refer to low and high profitability.

Brand	Profit	alphabet
10550	High	A
10550	Low	a
29585	High	B
29585	Low	b
13002	High	C
13002	Low	c

Fig. 2. Mapping Table of Combined strings

Compared to conventional methods of summarizing purchase history data, the proposed fusion method has the following advantages:

1) The sequence of the alphabet in the string pattern retains the time information related to purchase. Thus, the transformed data capture the change in customer behavior over time.

2) Users do not need to rely on ad hoc rules to extract raw purchase history data into analyzable form. The data transformation is accomplished via a mapping table. Thus it is easy to transform the purchase history data into analyzable format.

2.2 Algorithm for string pattern analysis

We provide an outline of BONSAI and the proposed algorithm. Suppose for simplicity that the original data set consists of a single numerical or categorical attribute together with time related quantity A_t. Assume that data are transformed into a string over the alphabet Γ (which corresponds to a purchase history of one customer in our case). Further, suppose that each string is labeled as *positive* or *negative*. Given a positive set of strings, *pos*, and a negative set of strings, *neg*, BONSAI creates a decision tree that attempts to accurately classify *pos* and *neg*. As a candidate set of regular patterns over an alphabet to be used at internal nodes of a decision tree, we enumerate regular patterns appearing in the strings of *pos* and *neg* and select those that appear more than a certain number of times. The maximum length of a regular pattern is usually fixed.

In order to generate a decision tree, we apply alphabet indexing to the set of strings. Let ϕ denote the alphabet indexing that maps Γ into another alphabet Γ' with smaller size. In our application that will be described in Section 3, the size of Γ' is fixed to a small constant, say two or three. When an alphabet indexing scheme is given, a decision tree can be constructed in a top-down manner as proposed by Quinlan [15–17]. We search for an alphabet indexing that has the highest classification accuracy by local search with multi-starts. Alphabet indexing can be viewed as information fusion that appropriately aggregate the original information into compact, meaningful forms. In order to accommodate general regular patterns instead of substrings, Hirao et al. [10] developed a branch-and-bound algorithm that incorporates an effective pruning method and a clever use of data structure. Our algorithm is based on the work of [10], but we had introduced several new features. We elaborate on specific details about these special features of the proposed algorithm:

a) The proposed algorithm allows for multiple sequences in constructing an accurate decision tree. This represents an important improvement over single-sequence methods because purchase history data often contain purchase records of multiple products from different categories. For example, in the analysis of such history, multiple sequencing allows the decomposition of data into multiple strings, each of which corresponds to a specific product category.

b) The proposed algorithm handles substrings which are of the form $\alpha\$$, where α is a substring, and the symbol $\$$ represents the termination of a sequence. Purchase patterns appearing at the end of the sequence of purchase history may have stronger predictive power on future purchase behavior. Therefore the ability to handle initiation and termination is important. This concept will be exemplified in Section 3. The proposed algorithm can also handle substring $\alpha\#(n)$, in which string α appears within the last n position within the sequence.

c) Another important new feature of the proposed algorithm is a region-based BONSAI, or Region-BONSAI. This extension to BONSAI is based upon the work of [14]. In this work, the authors proposed a classification algorithm – the majority weighted decision – which aims to produce simple classification rules. Technical details of the method are given in the next subsection. In order to distinguish decision-tree based BONSAI of [3,19] from Region-BONSAI, we adopt the term Tree-BONSAI for the decision-tree based BONSAI. Region-BONSAI uses the techniques of information fusion that make better decision by appropriately fusing several simple decision rules.

The computing time of our algorithm for both of Tree-BONSAI and Region-BONSAI is within a few seconds when the number of instances is less than a few hundred. However, the computing time rapidly increases when the number of instances increases. This phenomenon was also observed in [10]. The improvement of computing time for large-scale samples will be an important area for future research.

2.3 Region-BONSAI

In this subsection, we provide an overview of the scheme of weighted majority decision among region rules. We then explain how Region-BONSAI is developed based on the weighted majority decision.

Weighted majority decision among several region rules is originally proposed by Nakaya et al, [14]. They considered the classification problem of how to predict the values of a categorical attribute using other numerical attributes. To address classification problems, they used a weighted majority decision among region rules in which the component rules were relatively powerful and visually understandable. To improve the power of classifiers, they employed a strategy called decision by voting. As the term suggests, this technique makes the final judgment by a majority decision among component voters. Accuracy is obviously a requisite element for prediction of attribute values. However, to achieve a high predictive accuracy often requires predictors of considerable size. If a majority decision among a large number of voters is used, then the algorithm may suffer from poor rule readability problems. Thus, a small set of voters with high prediction accuracy is desired.

We shall explain in more details the framework of weighted majority decision. Figure 3 shows how two-dimensional (2-D) region is constructed by a pair of conditional attributes and a target attribute that are binary: 1(positive), 0(negative). A 2-D rule for two numerical attributes A and B is defined according to the following procedure. Suppose the domains of A and B are appropriately discretized, then the entire region R in two dimensions of A and B can be viewed as a rectangle consisting of $|A| \cdot |B|$ cells, where $|A|$ and $|B|$ are respectively the domain sizes of A and B after discretization (see Figure 3).

A 2-D rule determines a record with unknown target value as positive if values for a pair of attributes A and B are inside R_0 and as negative otherwise. Thus, to obtain a 2-D region rule, we identify the subset R_0 of the rectangle R such that splitting R into R_0 and $R - R_0 = (R_1)$ results in the largest information gain. In [14], the information gain is measured by entropy gain. From the viewpoint of interpretability of the region rule, it is useful to restrict the type of region R_0 to either rectangle, x-monotone region, or rectilinear convex region [7,24].

In [14], the weighted majority rule is computed as follows. If the number of conditional attributes is m, ${}_mC_2$ region rules are calculated. Among those region rules, those with k highest entropy gains are used for voting. They are called *voters*. The optimal procedure for determining the threshold value k

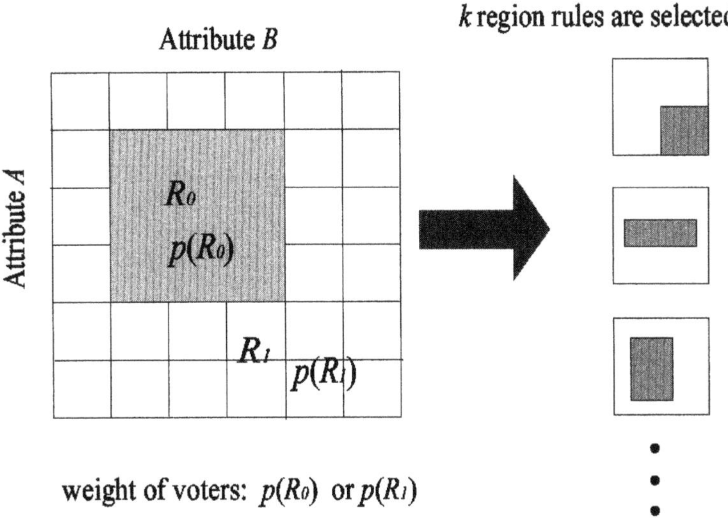

Fig. 3. 2-D region rule and weighted majority decision

will be described later. A nonnegative weight between 0 and 1 that is attached to each voter is determined by the prediction accuracy of the voter for the training dataset. For the set of tuples which are inside (outside) the region R_0, the ratio, $p(R_0)$ ($p(R_1)$), of tuples whose target attribute is positive is calculated. The ratio $p(R_0)$ ($p(R_1)$) is then used as the weight for where the value of the target attribute is positive. Specifically, for a tuple whose target value is unknown, if the values for a pair of attributes are inside R_0 (resp. R_1), the tuple receives the weight $p(R_0)$ (resp. $p(R_1)$). The summation of the weights of the k selected voters indicates the degree of plausibility that the target attribute of the tuple is positive. If the summation of weights is greater than a threshold $k/2$, the majority decision predicts that the target attribute of the tuple is positive.

In order to clarify how the majority voting works, we shall give the following illustrative example. Let us consider the problem of predicting whether a new customer becomes a loyal customer in the future based on three attributes; average profit per visit and the number of visits for few months from the first arrival as well as the customer age (see Section 3.2 for the details about loyal customer analysis). The target attribute is positive or negative, i.e., loyal customer or not loyal customer. Suppose each numeric attribute is appropriately discretized into five categories. There are three combinations of choosing two attributes from among three. For each pair of attributes, we generate the best region rule as illustrated in Fig. 4 according to the method described above. Here the shape of the region is rectilinear convex. Each region rule is associated with the weights $p(R_0)$ and $p(R_1)$ shown in Fig. 4. Let us assume all three voters are selected. For a tuple whose target value is unknown as in Fig. 4, the values for every pair of attributes are inside R_0, and the sum of weights is 2.0 which exceeds the threshold 1.5. Thus, the tuple is predicted as a loyal customer.

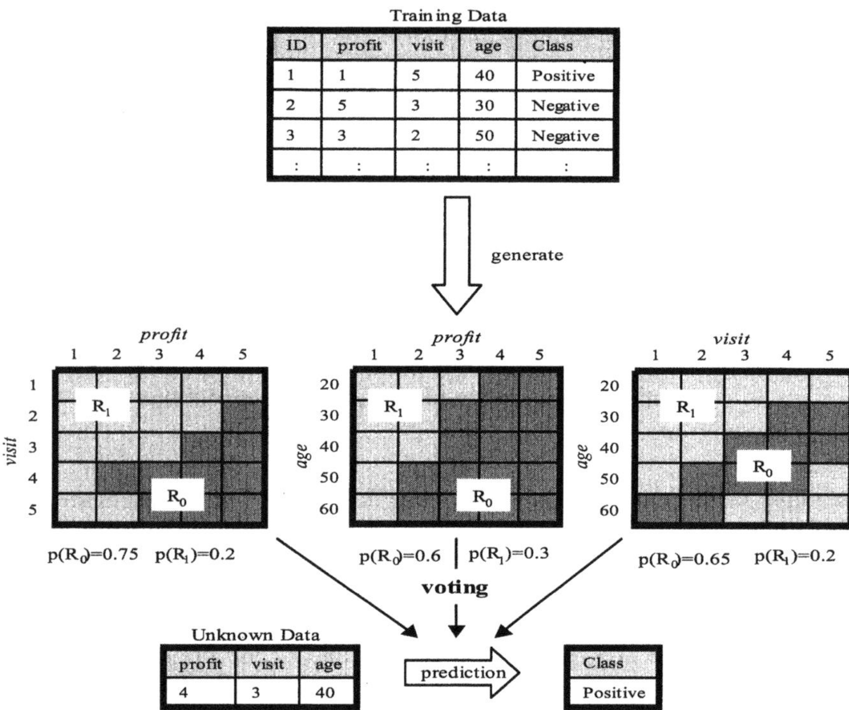

Fig. 4. Illustrative example for majority decision based on region rules.

In a recent paper [11], the method of [14] is improved in the following way:

1. A better gain index is used. The use of entropy gain in calculating region rules seems to lack strong theoretical basis. It was theoretically shown in [11] that instead of entropy gain, the use of Gini index always produces the better prediction accuracy.
   Rigorous definitions of entropy gain and Gini index are given as follows. Let $s(R)$ denote the number of records (customers in our application) in the database under consideration, and let $s(R_0)$ and $s(R_1)$ denote the number of records that fall into regions R_0 and R_1, respectively. Let $h(S)$ denote the number of records among those in region S such that the target attribute is "positive". Let

$$p = \frac{h(R)}{s(R)}, p_0 = \frac{h(R_0)}{s(R_0)}, p_1 = \frac{h(R_1)}{s(R_1)}.$$

$$n = \frac{s(R) - h(R)}{s(R)}, n_0 = \frac{s(R_0) - h(R_0)}{s(R_0)}, n_1 = \frac{s(R_1) - h(R_1)}{s(R_1)}.$$

Then, as defined in [15,16], the entropy gain $\mathrm{Ent}(R_0, R_1)$ with respect to region splitting of R into R_0 and R_1 is given by

$$\mathrm{Ent}(R_0, R_1) = -(p \log_2 p + n \log_2 n) + \frac{s(R_0)}{s(R)}(p_0 \log_2 p_0 + n_0 \log_2 n_0) + \frac{s(R_1)}{s(R)}(p_1 \log_2 p_1 + n_1 \log_2 n_1).$$

Gini index $Gini(R_0, R_1)$ with respect to region splitting of R into R_0 and R_1 is defined as

$$Gini(R_0, R_1) = (1 - p^2 - n^2) - \frac{s(R_0)}{s(R)}(1 - p_0^2 - n_0^2) - \frac{s(R_1)}{s(R)}(1 - p_1^2 - n_1^2).$$

It can be shown that maximizing the expected prediction accuracy is equivalent to maximizing the expected weight that a positive record receives, which is given by

$$h(R_0)p_0 + (s(R_0) - h(R_0))n_0 + h(R_1)p_1 + (s(R_1) - h(R_1))n_1. \quad (1)$$

From this expression, it can be easily derived that the maximizing (1) is equivalent to maximizing $Gini(R_0, R_1)$ (see [11] for the details).

2. The selection procedure for threshold of majority is optimized. Although the term majority literally means $k/2$, it is not clear that $k/2$ is optimal. We optimize the threshold to maximize the prediction accuracy in the training dataset.
3. The procedure for determining the number of voters is improved. The number of voters is optimized to maximize the prediction accuracy in the training dataset.

In the following application, we use region-BONSAI as follows: from a set of strings corresponding to purchase history of customers, we first generate a candidate set of regular patterns. For each of regular patterns (say, α), we generate a binary attribute such that it takes the value "positive" or "negative" depending on whether a symbol string contains α or not. A candidate set of 2-D region rules are obtained for all pairs of binary attributes generated. For a pair of binary attributes A and B, the entire region of $A \times B$ can be viewed as 2×2 table. Figure 5 shows that there are 14 nontrivial ways of splitting the entire region R into R_0 and R_1. Empty region or entire region ((15) and (16) in the figure) are not meaningful in the present context.

Notice that the cases of (11) through (14) are viewed as one-dimensional (1-D) rule. Since we separately treat 1-D rules and 2-D rules in Region-BONSAI, these cases are excluded from consideration. Therefore, we only consider the 10 possible types of regions (1) through (10) in Figure 5.

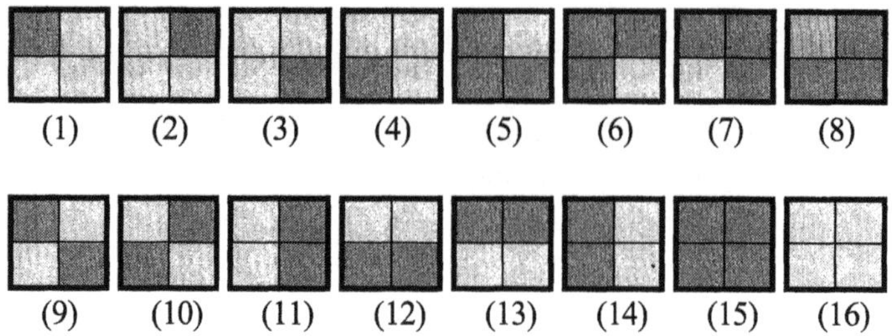

Fig. 5. 16 ways of region splitting.

3 Cases for Analysis

We applied the proposed algorithm to three different situations: brand switching analysis on baby diaper, loyal customer analysis, and customer attrition analysis. We demonstrate that interesting rules can be discovered and utilized for creating effective marketing strategy. All datasets have been obtained from purchase history accumulated in a drugstore chain in Japan [9]. The drugstore chain has approximately three million membership customers. It had started gathering purchase histories of each of its membership customers since the late 1980's.

In our analysis, we applied both Tree-BONSAI and Region-BONSAI. For regular pattern we use in generating decision rules for both of Tree-BONSAI and Region-BONSAI, we applied the method to the following three cases:

1) substrings.
   In this case, only substrings are used.
2) substrings+\$,#.
   In addition to substrings, substrings expressed as $\alpha\$$ $\alpha\#()$ are used in order to enhance predictive power on future purchase behavior as mentioned in Section 2.2.
3) all regular patterns.
   In this case, we do impose any restriction about the form of regular pattern.

3.1 Brand switching analysis on baby diaper

In this experiment, we consider the brand switching behavior of baby diaper users. In the Japanese market of baby diaper, there are seven major brands. Baby diapers are classified according to their size. We are interested in sizes M and L. We selected one brand of diaper, hereafter called the target brand, against which the remaining brands will be compared. Our objective is to predict which customers will become loyal users of the target brand in the category of size L by examining purchase pattern of baby diapers of size M.

The analysis is important because diaper users tend to evolve over their lifecycles. For example, manufacturers are interested in factors that affect brand choice behavior: how and when customers form their preference. Knowing who will likely to use what and when the consumer will use the product provides both manufacturers and retailers marketing opportunities for timely and customized offers.

The market share of the selected target brand in sizes M and L was about 24% in 1999. Most buyers of baby diapers were young women. They were keen on the quality of the products and often had their own preference of brands. The dataset used in the experiment was extracted according to the following procedure.

From the purchase history accumulated during the period from 1996 to 1999 from about 100 drugstores, we selected customers who purchased baby diapers of size M at least four times, and size L at least five times after switching from size M to L. Among these customers, we selected 918 customers who were loyal to the target brand of size L. A customer is said to be loyal to brand X of size L if at least 66.7% of baby diapers of size L purchased are from brand X. These customers were treated as positive instances in subsequent analyses. In general, let $amount_X$ denote the amount of brand X baby diapers purchased by the customer, and $amount_{all}$ denote the total amount of baby diapers purchased by the customer. The loyalty index of the customer to brand X is defined as $amount_X / amount_{all}$. For negative instances, we selected 918 customers whose loyalty index for the target brand was less than or equal to 33.3%. As a result we included 1,836 customers in the data set.

For each included customer, we transformed purchase history to string pattern. The seven competing baby diapers were coded as A,B,C,D,E,F and G, while the remaining miscellaneous brands were aggregated as a single brand H. In our analysis, brand A is selected as a target brand. Alphabets were then concatenated according to the order of purchase. Table 1 shows a sample of the transformed data.

Table 1. A sample data set of brand switching analysis

Customer ID	brand pattern	Class
a	DDAAA	Loyal
b	BABCAAAAA	Loyal
c	ABBBAB	NotLoyal
d	FCEABCCC	NotLoyal

Using this transformed data set, we applied Tree-BONSAI and Region-BONSAI for three types of regular pattern. The results are summarized in Table 2. For both algorithms, a 5-fold cross-validation procedure was imple-

Table 2. Accuracy and size of decision rules obtained by BONSAI. "Size" denotes the average number of leaves of decision trees for Tree-BONSAI and the number of voters for Region-BONSAI.

	Tree-BONSAI		Region-BONSAI	
	accuracy	size	accuracy	size
substrings	84.8%	2	85.6%	19
substrings+$, #	86.5%	2	86.7%	9
all regular patterns	86.4%	2	86.5%	3

mented. Values shown in Table 2 are averages of results from 5-fold cross-validation. The conventional way of implementing 5-fold cross-validation, the ratio of data size of training data and test data is set to 4. Namely, the whole data set is partitioned into five subsets of equal size, and generate five different training dataset each of which is a union of four from among five subsets. Prediction models are then built based on Tree- and Region-BONSAI for the union of four subsets and test the accuracy of the model using the rest of the data. However, in our case, taking into consideration the computational inefficiency of BONSAI, we execute 5-fold cross-validation in a different way: after partitioning the whole data set into five subsets of equal size, we choose each subset as a training set while the rest of the data is treated as a test data (i.e., the ratio of the sizes of training data and test data is set to 1/4).

Region-BONSAI exhibits slightly higher accuracy for any of three types of regular patterns. In addition, as we have expected, the prediction accuracy was increased when we use substrings with $ and #. The improvement of the accuracy may be due to the feature that can take into account the position of the substring.

Figure 6 shows one of five tress obtained by using substrings as regular patterns in Tree-BONSAI. Rectangular box represents the conditional node in which '111' is used to classify the customer as being loyal or not loyal to brand A at size L. Namely, if the purchase history of a customer encoded by an alphabet indexing shown at the left of the figure contains '111' as a substring, she/he is classified as "loyal", and otherwise "not loyal". Alphabet indexing encodes the target brand A as 1 and the other seven brands as 0. The number in parentheses refers to the number of customers who match or not match the condition. The indexing indicates that purchase of target brand in size M has strong influence on the loyalty of the target brand in size L. The rule states that when a customer purchased baby diapers of size M from brand A three consecutive times, she/he would likely to become a loyal customer of size L of brand A.

Among five decision tress obtained, two of them are as shown in Fig. 6 while the other three use the substring '11' in the unique conditional node. Thus, it can be observed that purchasing diapers of brand A at two or more consecutive purchase opportunities is strongly related to the loyalty of the target brand in size L.

Notice that decision rules obtained by Tree-BONSAI are interpretable by practitioners, a fact that might not have been generally acknowledged.

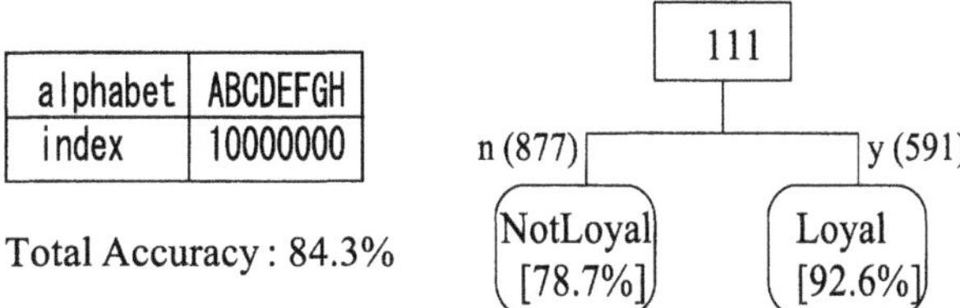

Fig. 6. A decision tree by Tree-BONSAI. '111' in the rectangle is the substring used for classification.

For the case of substrings+$,# of regular patterns Tree-BONSAI also produced very simple trees. Figure 7 shows one of such trees obtained in 5-fold cross validation. Selected alphabet indexing is identical to the previous one. The rule uses only one conditional node (i.e., '11#(3)'), and it states that if a customer purchased baby diaper of brand A at size M two consecutive times that occurred at the last three occasions, she/he would likely to become loyal to size L of brand A. This is an example that the position of the alphabet can also has a strong influence on prediction. Thus, the new feature of our method such that the position of the alphabet can be taken into account helps to generate an interesting rule about customer purchase behavior.

Among five decision trees, three of them are the one shown in Figure 7 while the other two uses the substring '111#(5)' in the unique conditional node.

For the case of general regular patterns, all of five decision trees have one conditional node. Regular patterns used are '1*11$', '1*11*1$', '11*1$'(used twice), '1*1*1$'. As in the case of substrings and substrings+$,#, it is observed that the purchase at the last purchase opportunity and consecutive purchases of brand A at size M have a strong influence on loyalty of brand A at size L.

We shall briefly mention the results by Region-BONSAI. In general, the rules obtained are more difficult to interpret than those by Tree-BONSAI. However, in order to exemplify the output of Region-BONSAI, we shall show a set of 2-D rules obtained for the case of substrings+$,# in one of 5-cross validation. It consists of three 2-D rules illustrated in Figure 8. Dark gray

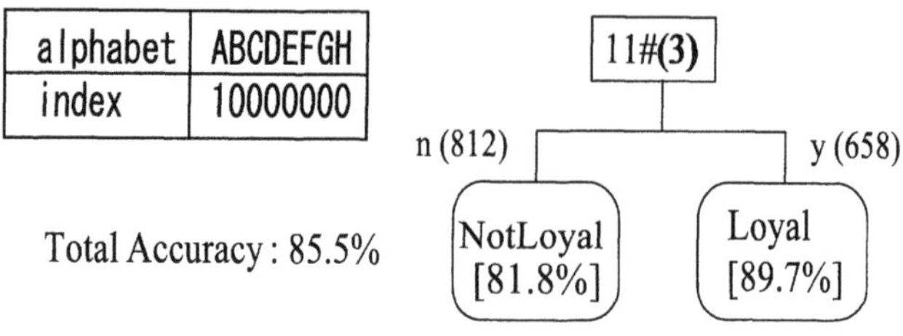

Fig. 7. A decision tree by Tree-BONSAI with suffix substrings

alphabet	ABCDEFGH
index	10000000

Accuracy: 85.3%		"00$" Yes	"00$" No
"11"	Yes	8/17	142/154
	No	27/183	7/14

(a)

Accuracy: 85.6%		"0$" Yes	"0$" No
"00#(4)"	Yes	38/207	24/31
	No	9/12	113/118

(b)

Accuracy: 85.6%		"1$" Yes	"1$" No
"00#(4)"	Yes	24/31	38/207
	No	113/118	9/12

(c)

Fig. 8. Three 2-D rules by Region BONSAI with suffix substrings

regions indicate customers who are predicted as being loyal to brand A of size L, and light gray regions indicate customers as being not loyal. For instance, rule in Figure 8(a) reads that if a customer purchased baby diapers of size M from brand A two consecutive times and also she/he bought diapers of size M from brand A at least once at the last two purchase opportunities, she/he would likely to become a loyal customer of size L of brand A. The string '00#(4)' used in rules (b) and (c) means that a customer purchased baby diapers of size M from brands other than A two consecutive times during the last four occasions. The rule in Figure 8(b) states that if the purchase history encoded as a binary string based on the alphabet indexing shown in Figure 8 contains '00#(4)' as well as '0$'(i.e., this means that the last purchased brand is not A), then the customer would not likely to become a loyal customer of size L of brand A. The rule in Figure 8(c) can be similarly interpreted. Similarly to Tree-BONSAI with substrings+$,#, it is observed that the purchase at the last few purchase opportunities have a strong influence on loyalty of brand A at size L.

3.2 Loyal Customer Analysis

This subsection concerns the analysis of loyal customers. In the brand switching analysis we transformed purchase history data into string patterns on the categorical attributes of a brand. The present experiment on loyal customer demonstrates string pattern analysis using numerical attributes.

In [13,12], a data mining technique was developed to identify potential high-value new visitors (called *loyal customers*) to the store at an early stage. The successful identification of potential new loyal customers early on is important because the company can use this information to establish a close relationship to this selected group of customers, thus building a tighter bond and reducing the chance that they will leave [5]. In the study of [13,12], a loyal customer is defined in terms of two dimensions; profitability per visit and frequency of visit. Each of these variables is categorized into five classes so that the size of each class is approximately the same. Such categorization used therein is illustrated in Table 3.

Table 3. Categorization of variables

	First dimension	Second dimension
class	Profit/visit (in yen)	Freq. of visits in 1998
5	566 or more	13 or more
4	315 to 565	7 to12
3	170 to 314	4 to 6
2	41 to 169	2 to 3
1	40 or less	1

Loyal customers are defined to be those who score one of (5, 5), (5, 4) or (4, 5) in these two dimensions. Accordingly, the loyalty status can only be confirmed one year after the first visit. Among 114,069 customers who became members in 1998, about 10% customers are classified as loyal. Such customers generated a disproportionate 52.5% of profit share and 38.4% of revenue share, exemplifying the strategic importance of the ability to manage relationship with this class of customers. Using profitability, frequency of visit[13,12], and other attributes for the first two, three or four months, [13](resp. [12]) showed that the algorithm C5.0 (resp. neural networks) exhibited satisfactory prediction ability for the identification of loyal customers.

In this experiment, we selected 16,902 customers among those used in [13,12] so that the numbers of loyal and non-loyal customers are equal. Our goal is to discover a simple rule to predict loyal customers using the data of the first 13 weeks. In our experiment, we use two sequences as conditional attributes. The first is a sequence of profitability per visit (PPV) encoded as a string of symbols $\{1, 2, 3, 4, 5\}$, which correspond to the five classes shown in Table 3. The length of the string is equal to the number of visits within the first 13 weeks, and the i-th symbol represents the class of profitability at the i-th visit. The second is a sequence of symbols $\{0, 1\}$ of length 13 that represents visit pattern such that the symbol 0 or 1 at the i-th position stand for non-visit or visit in the i-th week, respectively. Table 4 shows a sample of visiting patterns.

Table 4. Categorization of profitability

Customer ID	Profit Pattern	Visiting Pattern	Class
a	5343454	1100010010111	Loyal
b	345554	1001100001101	Loyal
c	435	0010000010001	Not Loyal
d	12111221	1011011001011	Not Loyal

In order to eliminate meaningless indexings, when selecting the best alphabet indexing for a sequence of profitability per visit, we limit alphabet indexing searching by the following procedure. Let Γ' be the set of alphabets into which an alphabet indexing ϕ transforms $\{1, 2, 3, 4, 5\}$. Then for any $a \in \Gamma'$, the set $\{i \in \{1, 2, 3, 4, 5\} \mid a = \phi(i)\}$ consists of consecutive integers. For example, if $\Gamma' = \{a, b\}$, ϕ should satisfy that there exists k with $1 \leq k \leq 4$ such that $\phi(1) = a, \ldots, \phi(k) = a$ and $\phi(k+1) = b, \ldots, \phi(5) = b$.

As in the previous case, we implemented 5-fold cross-validation for Tree- and Region-BONSAI for three types of regular patterns. The ratio of the size of training data and that of test data is set to 1/4.

Table 5 summarizes the results of the experiment. For each row in the table, a number denotes the average of the five tests of 5-fold cross-validation.

Table 5. Accuracy and size of decision rules obtained by Tree- and Region-BONSAI. The sizes for Tree-BONSAI and Region-BONSAI denote the number of leaves of decision trees and the number of voters used, respectively.

	Tree-BONSAI		Region-BONSAI	
	accuracy	size	accuracy	size
substrings	69.6%	2	70.6%	15
substrings+$,#	69.6%	2	70.3%	7
all regular patterns	70.8%	2	70.8%	11

It is observed from the table that Region-BONSAI enjoyed a slightly better performance. Among three types of regular patterns, that of all regular patterns was the best. The average number of leaves of decision trees obtained by 5-fold cross-validation tests was 2. In other words, the decision rule of every tree consists of only one conditional attribute. When substrings were used in generating decision trees by Tree-BONSAI, all of the five trees used the substring "1" of profitability sequence as the unique conditional attribute (Figure 9). The underlying alphabet index encodes profitability classes 1,2,3,4 as 0, and 5 as 1. This tree only used the profitability sequence but not the visit pattern. The rule in Figure 9 states that if in at least one of visits profitability lies in the highest class (class 5), then the customer is likely to be a loyal customer.

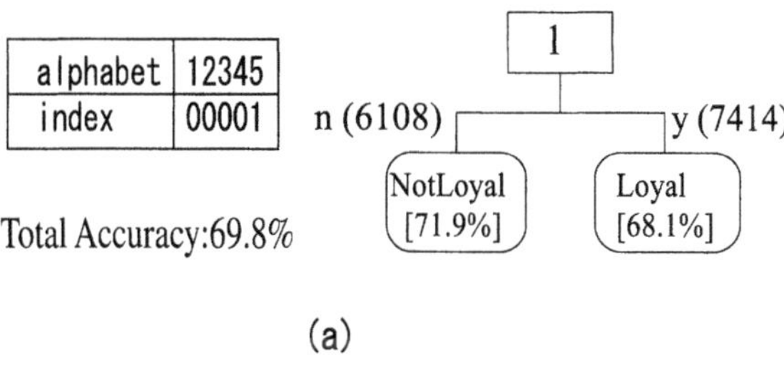

(a)

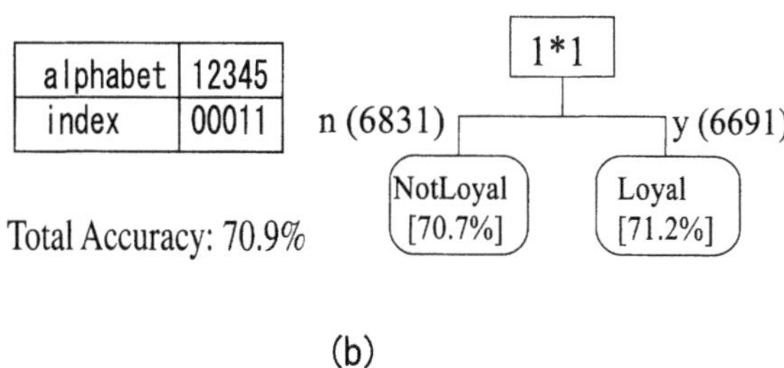

(b)

Fig. 9. Figure (a) shows a decision tree generated by Tree-BONSAI for loyal customer analysis when BONSAI uses only substrings. Figure (b) shows a decision tree generated by Tree-BONSAI for loyal customer analysis when BONSAI uses all regular patterns.

When substrings+$,# of regular patterns are used, five decision trees obtained are the same as in Figure 9(a). This implies that in the case of loyal customer analysis, substrings using $ and # did not contribute to the improvement of prediction accuracy.

When general regular patterns instead of substrings were used in generating decision trees by Tree-BONSAI, all of the five trees gave the same results (Figure 9(b)). The decision tree used the regular pattern "1*1" of PPV as the unique conditional attribute. The underlying alphabet index encodes profitability classes 1,2,3 as 0 and 4,5 as 1. The rule states that if in at least two visits profitability lies either in class 4 or 5, then the customer is likely to be a loyal customer. Both of the rules obtained are very simple. Because of their high interpretability, this kind of data-driven rules obtained by tools such as BONSAI should be useful to practitioners who work in loyalty programs. This indicates that the new feature of our method such that general regular patterns can be used to derive a decision tree is useful.

The details of results by Region-BONSAI are omitted in this case analysis. Although Table 5 shows that Region-BONSAI exhibits a slightly better accuracy than Tree-BONSAI, the interpretability of rules obtained by Region-BONSAI may not be as good as Tree-BONSAI as it uses 7 or more 2-D region rules as voters. In terms of simplicity and interpretability, Tree-BONSAI seems to have an edge over region-BONSAI in this case.

3.3 Customer Attrition Analysis

As in the previous subsection, the present experiment on customer attrition again demonstrates analysis on string patterns constructed from numerical attributes. The work is a continuation of [4] in which customer attrition model using string pattern analysis was developed.

While customer attrition cannot be entirely prevented, controlling the rate of attrition has become a critical business issue because of the high cost of acquiring new customers. If you were a store manager, you could not stay competitive unless you were able to gain insight into customer behavior, identify and solve potential problems before they caused customers to defect from your store.

In this case study, our objective is to use empirical analysis to assist management of a drugstore chain acquire critical information of customer behavior and evaluate likelihoods of customer defection in relation to purchase behavior. To carry out the analysis we used four years of purchase history data from the drugstore, and targeted those customers who visited the store at least one time from January to June in 1998. As a result, 7,074 customers are included in this study.

We distinguish between inactive("dead") and "active" customer by using the following criterion: a customer is defined as "dead" if he/she never visited the store for the 6-month period that ends on December 31, 1998, and he/she visited the store at least n times before June 30, 1988. Here n is a parameter to be chosen in our analysis in order to control the number of active and dead customers. We call the date of the customer's last visit his/her target date. On the other hand, an active customer is defined as follows. We select a target date from January to June in 1998 at random for each customer who were not classified as dead. Then, if the customer visited the store at least 6 times after the target date and he/she visited it at least n times before the target date. Notice that among 7,047 customers, there are those who are not classified as either active or dead.

In our experiment, we typically choose $n = 4, 6, 8, 10$ and 15. Table 6 shows the number of active and dead customers for each value of n. To build the training set, we randomly selected 400 customers, 200 from each group.

Table 6. The number of active or dead customers by n.

n	active	dead
4	3424	701
6	3320	596
8	3224	480
10	3096	402
15	2863	250

Our goal is to discover simple rules for identifying dead customers. We used two string patterns as conditional attributes: idle days between visits (IDBV) and profitability per visit (PPV). IDBV and PPV corresponding to n visits preceding the target date are encoded as a string of symbols from the set $\{1, 2, 3, 4, 5\}$, the elements of which correspond to the five classes shown in Table 7. Notice that thresholds to classify PPV into five classes are different from those in Table 3.

The length of the string for IDBV is equal to $n - 1$, while that for PPV is equal to n. The i-th symbol of IDBV string represents the class of IDBV between the i-th and $i+1$-th visit. The i-th symbol of PPV string represents the class of PPV at the i-th visit. Table 8 shows a sample of the data set.

Table 7. Alphabet and range of IDBV and PPV

alphabet	IDBV	PPV
1	$1 \sim 7$	~ 17
2	$8 \sim 14$	$18 \sim 172$
3	$15 \sim 24$	$173 \sim 378$
4	$25 \sim 43$	$379 \sim 759$
5	$44 \sim$	$760 \sim$

Table 8. Sample data set

Customer ID	IDBV	PPV	Class
a	324544	112213	dead
b	114555	211543	dead
c	231131	123342	active
d	421412	221121	active

We performed experiments using both Tree- and Region-BONSAI. For both algorithms, a 5-fold cross-validation procedure was implemented. The procedure was applied to samples formed from the five values of n. Figure 10 shows the rate of accuracy for value of n. To our surprise, the rate of accuracy is highest when n, the pattern length, is 6. When n increases beyond this value, the accuracy decreases with increasing value of n. We conjecture that data from 4 visits do not provide sufficient information for accurate prediction, while data from more than 8 times of visits contain more information about actives but less about deads, and as a result negatively affects the classification accuracy. In brief, the preceding 6 visits from the target date contains optimal information for predicting whether a customer is dead.

Table 7 summarizes the results of the experiments with $n = 6$. We have tested three cases of regular pattern. In the cases of substrings and all regular patterns we have tested only the case in which both IDBV and PPV strings are used, while in the case of string+$, #, we have also tested the cases in which only PPV or IDBV is used. For each row in Table 9, a number denotes the average of five tests from the 5-fold cross-validation. Among three types of regular patterns, that of all regular patterns was the best. It is observed that in the case of customer attrition analysis, substrings using $ and # did not contribute to the improvement of prediction accuracy. It is also observed that for any of three types of regular patterns, Region-BONSAI exhibits better accuracy than Tree-BONSAI.

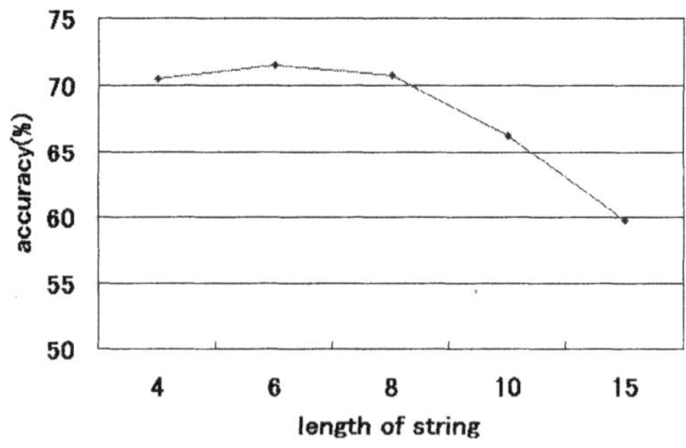

Fig. 10. Accuracy with respect to the change of the value of n.

Table 9. Accuracy and size of decision rules obtained by Tree- and Region-BONSAI.

		Tree-BONSAI		Region-BONSAI	
		accuracy	size	accuracy	size
substrings		72.5%	9	73.0%	9
substrings+$,#	IDBV	70.8%	4.8	71.8%	11
	PPV	57.5%	5.4	59.0%	5
	IDBV+PPV	71.5%	8.6	73.2%	19
all regular patterns		70.3%	6.8	72.8%	23

As easily seen from the table, PPV alone does not produce high accuracy. Thus, we examine the decision tree constructed by using IDBV. Figure 11 shows one such tree. The tree has five leaves and four internal nodes (conditional nodes). Some conditions overlap. For example, the second level condition '111$' contains the top level condition '11'. We can read the tree as follows:

a) If string pattern does not contain '11' then customer is active.
b) If string pattern contains '111$' or '1111' then customer is dead.
c) If string pattern contains '11' and '00' then customer is active.
d) Else customer is dead.

The underlying alphabet indexing encodes IDBV classes 1,2,3 as 0 and 4,5 as 1, i.e., index 0 means IDBV is 24 days or less (short IDBV) and index 1 means IDBV is 25 days or more (long IDBV). In summary, customers who have two or more consecutive long IDBV in the latest 6 visits are likely to become dead customers, with the exception of customers who have two consecutive short IDBV.

Although PPV did not seem to be a good predictor (accuracy is 57.5%), but it did contribute somewhat when used with IDBV. However, Tree-BONSAI generated complicated trees with a size of 8.6 on average. Under such circumstances, it seems more appropriate to use rules generated by Region-BONSAI, which are more interpretable.

Let us elaborate on specific details of the model building process based on Region-BONSAI using both IDBV and PPV, of which the prediction accuracy was highest (Table 9). First, for each of IDBV and PPV, we generate thirty regular patterns over corresponding alphabets with high classification ability. They are used as 1-D rules. On top of them, we generate 7,440($= 60 \cdot 59/2$) 2-D rules for all pairs of candidate regular patterns. Thus, we have a total of 7,500 region rules. Among them, those of k highest prediction accuracy are selected as voters. Figure 12 illustrates how the overall prediction accuracy of weighted majority decision changes as k changes from 1 to 29. At $k = 19$, the prediction accuracy reaches the highest value, 73.2% while the highest accuracy and the lowest accuracy among those selected 19 voters are 71.3% and 65.3%,

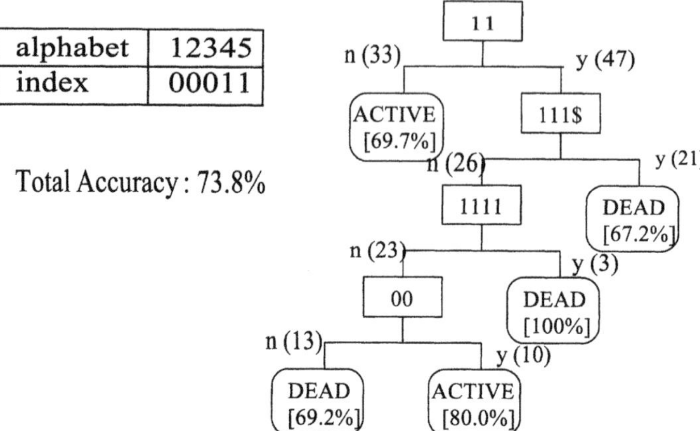

Fig. 11. Decision Tree on IDBV. Rectangular box represents conditional node, and the string in the box such as 11 in the topmost rectangle represents the substring used for classification.

respectively. Thus, k is set to 19 in the subsequent experiments. We observe here that the weighted majority decision contributes to the improvement of prediction accuracy.

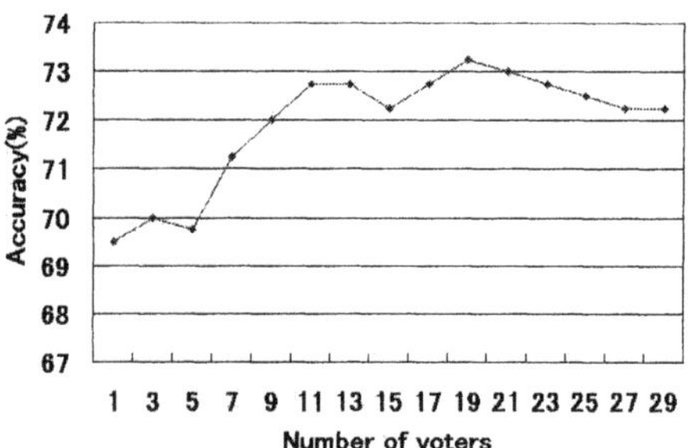

Fig. 12. The effect of the number of voters on the prediction accuracy.

We optimize the threshold value for the weight used in weighted majority decision with which target attribute of test data is predicted to be "positive" or "negative". For a training data set of size 320, we plot the weights that each data has collected from voters as illustrated in Figure 13. We choose the threshold that maximizes the prediction accuracy for the training data. In this case, 8.0 is the optimized threshold.

Here we show a 2-D region rule with the highest accuracy, and how the rule is interpreted (see Figure 14). In Figure 14, alphabet index encodes PPV classes 1,2 as 0 and 3,4,5 as 1, that is index 0 means PPV is 172 yen or less

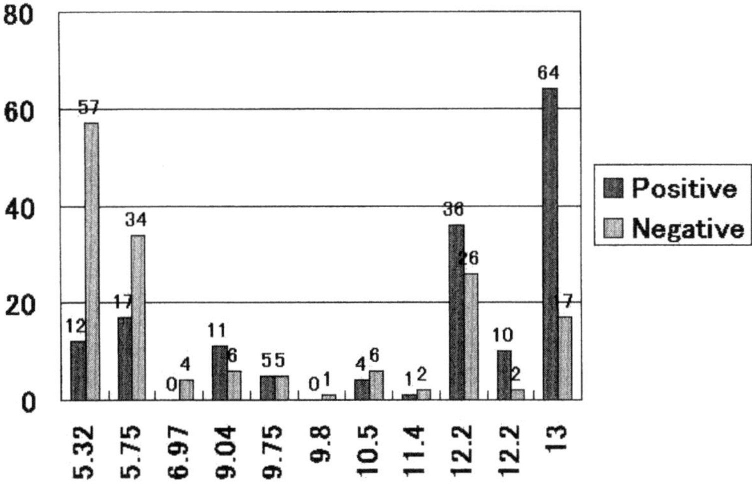

Fig. 13. The effect of threshold on the prediction accuracy.

(low PPV) and index 1 means 173 yen or more (high PPV), while IDBV is encoded in the same way as in the decision tree of Figure 11. The rule in Figure 14 depicts the conditions of IDBV as columns, and PPV as rows. Dark gray regions indicate customers who are predicted as being dead, and light gray regions indicate customers as being active.

Let us first consider the customers who do not have two consecutive long IDBV (the right column of the region). The rule provides the similar interpretation as that from decision tree. The rules states that, regardless of PPV, if a string pattern does not contain two consecutive long IDBV, then the customer is active.

For those with two consecutive long IDBV, one might be led to believe that they would not return, since they did not frequently visit the store. Contrary to this intuition, the rule tells us that if they have three consecutive low PPV, then they are active. Here we offer a plausible explanation. These customers visit the store only during bargain sales. Stores in Japan usually have bargain sales once a month. Bargain hunters visit the store once a month (long IDBV) to purchase competitively priced products (low PPV). They tend to continue to visit the store (stay active) as bargain hunters.

4 Conclusion

In this chapter, we developed a method for predicting future purchase behavior from customer purchase history by combining information fusion techniques with a data mining tool – string pattern analysis originally developed

	IDBV	PPV
alphabet	12345	12345
index	00011	00111

Accuracy: 71.6%		IDBV "11"	
		Yes	No
PPV "000$"	Yes	9/15	18/24
	No	37/143	96/138

Fig. 14. Rules by Region-BONSAI on IDBV and PPV

for genetic sequencing. The method for string pattern analysis we used was modified by adding new features so as to adapt to our purpose. We demonstrate how the combination of information fusion techniques with string pattern analysis can lead to simple and accurate models for predicting future purchase behavior. In particular, techniques of information fusion are used in three phases of the data mining process in our methods: (1) transformation from purchase history to string pattern, (2) optimal selection of alphabet indexing, and (3) majority voting in which a number of predictors are aggregated into a single one with appropriate weights. Through several real cases, we demonstrate how each phase contributes to the success of discovery of useful knowledge. It should be pointed out that the new features of our method such that general regular pattern can be used and the position of the alphabet can be taken into account contributed to the discovery of interesting rules concerning customer purchase behaviour.

References

1. Abidi M.A., Gonzalez, R.C. (Eds), (1992), Data Fusion in Robotics and machine Intelligence, Academic Press.
2. Agrawal R., T.Imielinski T., A., Swami A. (1993) Database Mining: A Performance Perspective, IEEE Transactions on Knowledge and Data Engineering, 5:914-925.
3. Arikawa S., Miyano S., Shinohara A., Kuhara S., Mukouchi Y., Shinohara T. (1993) A Machine Discovery From Amino Acid Sequences by Decision Trees over Regular Patterns. New Generation Computing 11:361-375.
4. Cheung S., Hamuro Y., Katoh N., Ip E.H., Yada K. (2001) Customer Attrition Analysis Using the History of Purchase Patterns. INFORMS International Hawaii Conference.
5. Chou P.B., Grossman E., Gunopulos D., Kamesam P. (2000) Identifying Prospective Customers, In Proc. KDD 2000, 447-456.
6. Fishman C. (1999) This is a Marketing Revolution, Fast Company, 206-218
7. Fukuda T., Morimoto Y., Morishita S., Tokuyama T. (1996). Data Mining Using Two-Dimensional Optimized Association Rules: Scheme, Algorithms, and Visualization. In Proc. of the ACM SIGMOD Conference on Management of Data, 13-23.

8. Hamuro Y., Kawata E., Katoh N., Yada K. (2001) A Machine Learning Algorithm for Analyzing String Patterns Helps to Discover Simple and Interpretable Business Rules from Purchase History, in Progresses in Discovery Science, State-of-the-Art Surveys, Lecture Notes of Computer Science, Springer-Verlag, 565-575.
9. Hamuro Y., Katoh N., Matsuda Y., Yada K. (1998) Mining Pharmacy Data Helps to Make Profits. Data Mining and Knowledge Discovery. 2(4):391-398
10. Hirao M., Hoshino H., Shinohara A., Takeda M, Arikawa S. (2000) A Practical Algorithm to Find the Best Subsequence Patterns. in Proc. of 3rd International Conference on Discovery Science. Lecture Notes on Artificial Intelligence 1967:141-154.
11. Horiguchi N., Yada K., Hamuro Y., Katoh N., Kambayashi Y. (2000) An Optimized Weighted Majority Decision. in Proc. of INFORMS-KORMS Seoul, 1663-1669.
12. Ip E., Johnson J., Yada K., Hamuro Y., Katoh N., Cheung S. (2002) A Neural Network Application to Identify High-Value Customers for a Large Retail Store in Japan, Chapter IV (pp. 55-69), Neural Networks for Business: Techniques and Applications. (Kate A. Smith, Jatinder N. D. Gupta eds.). Idea Group Publishing.
13. Ip E., Yada K., Hamuro Y., Katoh N. (2000) A Data Mining System for Managing Customer Relationship, in Proc. of the 2000 Americas Conference on Information Systems. 101-105
14. Nakaya A., Furukawa H., Morishita S. (1999) Weighted Majority Decision among Several Region Rules for Scientific Discovery, in Proc. of Second International Conference on Discovery Science, Lecture Notes on Artificial Intelligence 1721, Springer-Verlag, 17-29.
15. Quinlan J.R. (1986) Induction of Decision Trees, Machine Learning 1:81-106.
16. Quinlan J.R. (1993) C4.5: Programs for Machine Learning. Morgan Kaufman
17. Quinlan J.R. (1999) See5/C5.0, `http://www.rulequest.com`, Rulequest Research.
18. Piatetsky-Shapiro G, (Editor) (1991) Knowledge Discovery in Databases AAAI Press.
19. Shimozono S., Shinohara A., Shinohara T., Miyano S., Kuhara S., Arikawa S. (1994) Knowledge Acquisition from Amino Acid Sequences by Machine Learning System BONSAI. Trans. Information Processing Society of Japan 35:2009-2018.
20. Spangler W.E., May J.H., Vargas L.G. (1999) Choosing Data-mining Methods for Multiple Classification: Representational and Performance Measurement Implications for Decision Support. Journal of Management Information System 16(1):37-62.
21. Sung T.K., Chung H.M., Gray P. Special Section: Data Mining. Journal of Management Information System 16(1):11-16.
22. Torra, V. (2002) Information fusion in data mining. Chapter in this book.
23. Uthurusamy R., Fayyad U.M., Spangler S. (1991) Learning Useful Rules from Inconclusive Data, In [18], 141-157.
24. Yoda K., Fukuda T., Morimoto Y., Morishita S., Tokuyama T. (1997). Computing optimized rectilinear regions for association rules. in Proc. of International Conference on Knowledge Discovery and Data Mining, 96-103.

Ensembling Classification Systems by a Fuzzy Rule-Based System

Tomoharu Nakashima and Gaku Nakai

Osaka Prefecture University, Gakuen-cho 1-1, Sakai, 599-8531 Osaka, JAPAN

Abstract. In this chapter, we propose an ensembling method for pattern classification problems. In our ensembling method, two different types of fuzzy rule-based systems are used. One is for classifying input patterns. We refer to this type of fuzzy rule-based systems as fuzzy rule-based classification systems. The other type of fuzzy rule-based systems determines which classification systems are used for the classification. This type of fuzzy rule-based systems is referred to as fuzzy rule-based ensembling systems. Our ensembling method consists of one fuzzy rule-based ensembling system, several fuzzy rule-based classification systems, and a gating node that is used for final classification. An input pattern is presented to both types of fuzzy rule-based system. Each of the fuzzy rule-based classification systems determines which class the input pattern is from, and the fuzzy rule-based ensembling system assigns a credit of the classification to each fuzzy rule-based classification system. In the gating node, all the information is collected and the final classification of the input pattern is performed. In computer simulations, we examine the performance of our ensemble learning method on several real-world pattern classification problems. Simulation results show that the performance of our ensembling method is better than the best single fuzzy rule-based classification system. We also show the simulation results of our method on unseen patterns to see how well our method generalizes.

1 Introduction

Fuzzy rule-based systems have been applied mainly to control problems [1–3]. Recently fuzzy rule-based systems have also been applied to pattern classification problems. There are many approaches to the automatic generation of fuzzy if-then rules from numerical data for pattern classification problems. Genetic algorithms have also been used for generating fuzzy if-then rules for pattern classification [4–6].

It is generally said that putting several systems together produces better performance than the best single system. For example, Ueda and Nakano [7] analytically showed that the generalization error of averaged outputs from multiple function approximators is less than that of any single function approximator. For pattern classification problems, various ensembling methods have been proposed [8–13]. For example, Battiti and Colla [8] examined voting schemes such as a perfect unison and a majority rule for combining multiple neural networks classifiers. Hansen [13] theoretically and experimentally showed that the upper bound of misclassification by a neural network ensemble classifier with plurality voting is lower than a single neural network classifier. Cho and Kim [14–16]used fuzzy integrals for aggregating outputs from multiple neural networks. Ishibuchi et al.[17] examined the performance of different levels of voting for fuzzy rule-based classification systems such as a voting by multiple fuzzy if-then rules and a voting by multiple fuzzy rule-based classification systems.

In this chapter, we propose an ensembling method by using a fuzzy rule-based system (Fig. 1). In our proposed method there exist two different fuzzy rule-based systems. One is fuzzy rule-based classification systems which suggest the class of an input pattern, and the other is fuzzy rule-based ensembling systems which assign a weight value to each fuzzy rule-based classification system. A gating node uses the information about the suggested class for the input pattern and weight values of the fuzzy rule-based classification system to finally determine the class of the input pattern.

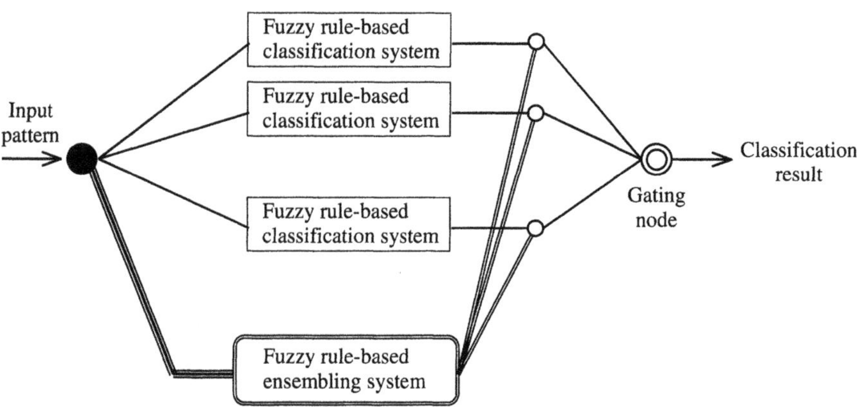

Fig. 1. Proposed ensembling system

In this chapter, we first describe an automatic generation procedure from given training data for designing a fuzzy rule-based classification system and a fuzzy inference procedure for classifying an input pattern. We also explain the learning procedure of the grade of certainty in order to improve the classification ability of the fuzzy rule-based classification system. Then, we explain our ensembling method that consists of several fuzzy rule-based classification systems, a single fuzzy rule-based ensembling system, and a gating node. In computer simulations, we examine the performance of the proposed method on several real-world pattern classification problems.

2 Fuzzy Rule-Based Classification System

2.1 Pattern Classification Problems

Various methods have been proposed for fuzzy classification [18]. Let us assume that our pattern classification problem is an n-dimensional problem with C classes. We also assume that we have m given training patterns $\boldsymbol{x}_p = (x_{p1}, x_{p2}, \ldots, x_{pn})$, $p = 1, 2, \ldots, m$. Without loss of generality, each attribute of the given training patterns is normalized into a unit interval $[0, 1]$. That is, the pattern space is n-dimensional unit hypercube $[0, 1]^n$ in our pattern classification problems.

In this study, we use fuzzy if-then rules of the following type in our fuzzy rule-based classification systems:

$$\text{Rule } R_j\text{: If } x_1 \text{ is } A_{j1} \text{ and } x_2 \text{ is } A_{j2} \text{ and } \ldots \text{and } x_n \text{ is } A_{jn} \\ \text{then Class } C_j \text{ with } CF_j,\ j=1,2,\ldots,N, \tag{1}$$

where R_j is the label of the j-th fuzzy if-then rule, $A_{j1}, \ldots, A_{jn}$ are antecedent fuzzy sets on the unit interval $[0, 1]$, C_j is the consequent class (i.e., one of the given C classes), CF_j is the grade of certainty of the fuzzy if-then rule R_j, and N is the total number of fuzzy if-then rules. As antecedent fuzzy sets, we use triangular fuzzy sets as in Fig. 2 where we show various partitions of a unit interval into a number of fuzzy sets.

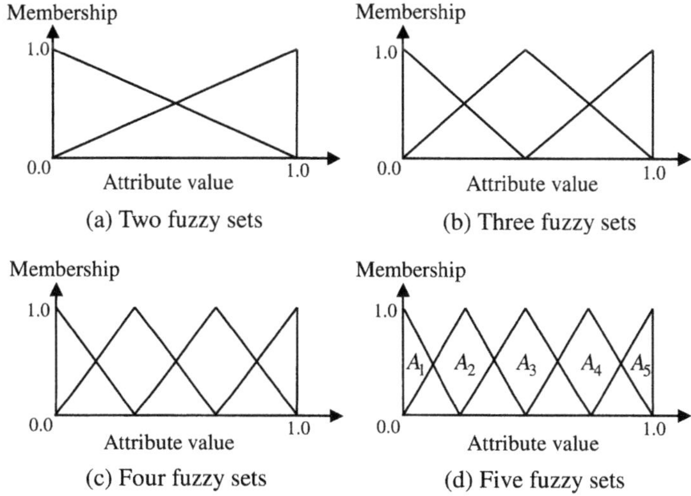

Fig. 2. An example of antecedent fuzzy sets

2.2 Generating Fuzzy If-Then Rules

In our fuzzy rule-based classification systems, we specify the consequent class and the grade of certainty of each fuzzy if-then rule from the given training patterns [19–21]. In [21], it is shown that the use of the grade of certainty in fuzzy if-then rules allows us to generate comprehensible fuzzy rule-based classification systems with high classification performance.

The consequent class C_j and the grade of certainty CF_j of fuzzy if-then rule are determined in the following manner:

[Generation Procedure of Fuzzy If-Then Rule]

1. Calculate $\beta_{\text{Class } h}(R_j)$ for Class h $(h = 1, \ldots, C)$ as

$$\beta_{\text{Class } h}(R_j) = \sum_{\boldsymbol{x}_p \in \text{Class } h} \mu_{j1}(x_{p1}) \cdot \ldots \cdot \mu_{jn}(x_{pn}), h = 1, 2, \ldots, C. \quad (2)$$

2. Find Class $\hat{h}$ that has the maximum value of $\beta_{\text{Class } h}(R_j)$:

$$\beta_{\text{Class } \hat{h}}(R_j) = \max\{\beta_{\text{Class } 1}(R_j), \beta_{\text{Class } 2}(R_j), \ldots, \beta_{\text{Class } C}(R_j)\}. \quad (3)$$

   If two or more classes take the maximum value, the consequent class C_j of the rule R_j can not be determined uniquely. In this case, specify C_j as $C_j = \phi$. If a single class takes the maximum value, let C_j be Class $\hat{h}$.
3. If a single class takes the maximum value of $\beta_{\text{Class } h}(R_j)$, the grade of certainty CF_j is determined as

$$CF_j = \frac{\beta_{\text{Class } \hat{h}}(R_j) - \bar{\beta}}{\sum \beta_{\text{Class } h}(R_j)}, \quad (4)$$

   where

$$\bar{\beta} = \frac{\sum_{h \neq \hat{h}} \beta_{\text{Class } h}(R_j)}{c - 1} \quad (5)$$

The number of fuzzy if-then rules in a fuzzy rule-based classification system is dependent on how each attribute is partitioned into fuzzy subsets. For example, when we divide each attribute into three fuzzy subsets in a ten-dimensional pattern classification problem, the total number of fuzzy if-then rules is $3^{10} = 59049$. This is what is called the curse of dimensionality. The grade of certainty CF_j can be adjusted by a learning alogrithm [22].

2.3 Fuzzy Reasoning

By the rule generation procedure in 2.2, we can generate N fuzzy if-then rules in (1). After both the consequent class C_j and the grade of certainty CF_j are determined for all the N fuzzy if-then rules, a new pattern $\boldsymbol{x}$ is classified by the following procedure [19]:

[Fuzzy reasoning procedure for classification]

1. Calculate $\alpha_{\text{Class } h}(\boldsymbol{x})$ for Class h, $j = 1, 2, \ldots, C$ as

$$\alpha_{\text{Class } h}(\boldsymbol{x}) = \max\{\boldsymbol{\mu}_j(\boldsymbol{x}) \cdot CF_j | C_j = \text{Class } h, h = 1, 2, \ldots, N\}, \quad h = 1, 2, \ldots, C, \tag{6}$$

   where

$$\boldsymbol{\mu}_j(\boldsymbol{x}) = \mu_{j1}(x_1) \cdot \ldots \cdot \mu_{jn}(x_n). \tag{7}$$

2. Find Class h_p^* that has the maximum value of $\alpha_{\text{Class } h}(\boldsymbol{x})$:

$$\alpha_{\text{Class } h_p^*}(\boldsymbol{x}) = \max\{\alpha_{\text{Class } 1}(\boldsymbol{x}), \ldots, \alpha_{\text{Class } C}(\boldsymbol{x})\}. \tag{8}$$

   If two or more classes take the maximum value, then the classification of $\boldsymbol{x}$ is rejected (i.e., x is left as an unclassifiable pattern), otherwise assign $\boldsymbol{x}$ to Class h_p^*.

2.4 Learning the Grade of Certainty

The performance of the fuzzy rule-based classification system can be improved by fine tuning the grade of certainty CF_j for each fuzzy if-then rule. Nozaki et al.[22] has proposed a learning algorithm of the grade of certainty CF_j. Each of the given training patterns is classified by a fuzzy rule-based classification system using the fuzzy reasoning procedure in Subsection 2.3. From the fuzzy reasoning procedure, we can see that only a single fuzzy if-then rule that has the maximum product of the compatibility grade and the grade of certainty is used for determining the class of an input pattern. That is, an input pattern $\boldsymbol{x}$ is classified by the fuzzy if-then rule R_j^* that satisfies the following relation:

$$\boldsymbol{\mu}_j^*(\boldsymbol{x}) \cdot CF_j^* = \max\{\boldsymbol{\mu}_j(\boldsymbol{x}) \cdot CF_j | j = 1, 2, \ldots, N\}, \tag{9}$$

where N is the total number of designing generated fuzzy if-then rules for a fuzzy rule-based classification system.

In the learning algorithm in [22], when $\boldsymbol{x}$ is correctly classified by the linguistic classification rule R_j^*, the grade of certainty CF_j^* of the fuzzy if-then rule R_j^* is increased as the reward of the correct classification as follows:

$$CF_j^* := CF_j^* + \eta_1 \cdot (1 - CF_j^*), \tag{10}$$

where η_1 is a positive learning rate for increasing the grade of certainty CF_j^*. On the contrary, when $\boldsymbol{x}$ is misclassified by the fuzzy if-then rule R_j^*, the grade of certainty CF_j^* is decreased as the punishment of the misclassification as follows:

$$CF_j^* := CF_j^* - \eta_2 \cdot CF_j^*, \tag{11}$$

where η_2 is a positive learning rate for decreasing the grade of certainty.

This learning algorithm is iterated until some prespecified stopping condition is satisfied. Usually, the number of iteration of the learning algorithm N_{learn} is used for the stopping condition.

2.5 Illustrative Example

In this subsection, we briefly explain how our fuzzy rule-based classification sytem works for an artificial two-dimensional classification problem. Then we also show that the performance of the fuzzy rule-based classification system can be improved by the learning algorithm of the grade of certainty. Let us assume that we have a two-class two-dimensional pattern classification problem with 15 training patterns from each class as in Fig. 3. We also assume that each attribute is divided into five fuzzy subsets (see Fig. 2 (d)). Thus the total number of generated fuzzy if-then rules is $5^2 = 25$.

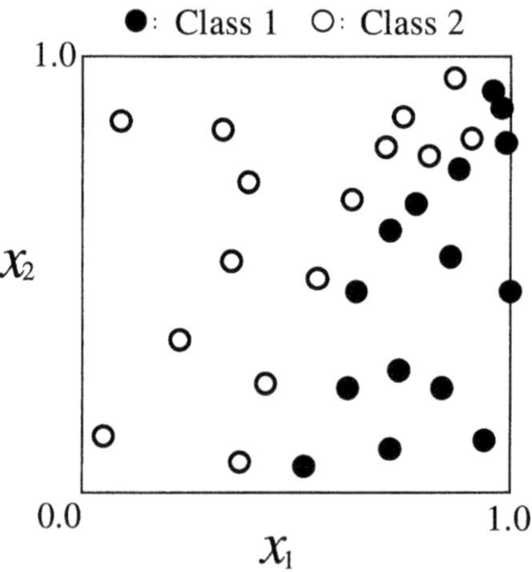

Fig. 3. A two-dimensional pattern classification problem

Since there are only two classes in this example, the grade of certainty CF_j of fuzzy if-then rule $R_j, j = 1, 2, \ldots, 25$ is calculated as follows:

$$CF_j = \frac{\beta_{\text{Class 1}}(R_j) - \beta_{\text{Class 2}}(R_j)}{\beta_{\text{Class 1}}(R_j) + \beta_{\text{Class 2}}(R_j)}, \quad j = 1, 2, \ldots, 25. \tag{12}$$

We show the generated fuzzy if-then rules for our illustrative two-dimensional problem in Table 1. In Table 1, $A_1, \ldots, A_5$ are the fuzzy sets specified in the Fig. 2 (d).

By using all the generated 25 rules, we can correctly classify 29 out of 30 training patterns. (i.e., 96.7% classification rate). In Fig. 4, we show the classification boundary between Class 1 and Class 2. From Fig. 4, we can see that the classification boundary is not linear but nonlinear. This is the main characteristic of the fuzzy rule-based classification system. The use of the grade of certainty is one of the important features of our fuzzy rule-based classification system (for more detail, see [21]).

Table 1. Generated fuzzy if-then rules (C_j and CF_j)

		x_1				
		A_1	A_2	A_3	A_4	A_5
x_2	A_5	Class 2 (1.00)	Class 2 (1.00)	Class 2 (1.00)	Class 2 (0.76)	Class 1 (0.38)
	A_4	Class 2 (1.00)	Class 2 (1.00)	Class 2 (0.92)	Class 2 (0.26)	Class 1 (0.45)
	A_3	Class 2 (1.00)	Class 2 (1.00)	Class 2 (0.51)	Class 1 (0.69)	Class 1 (1.00)
	A_2	Class 2 (1.00)	Class 2 (1.00)	Class 2 (0.01)	Class 1 (0.99)	Class 1 (1.00)
	A_1	Class 2 (1.00)	Class 2 (1.00)	Class 1 (0.39)	Class 1 (1.00)	Class 1 (1.00)

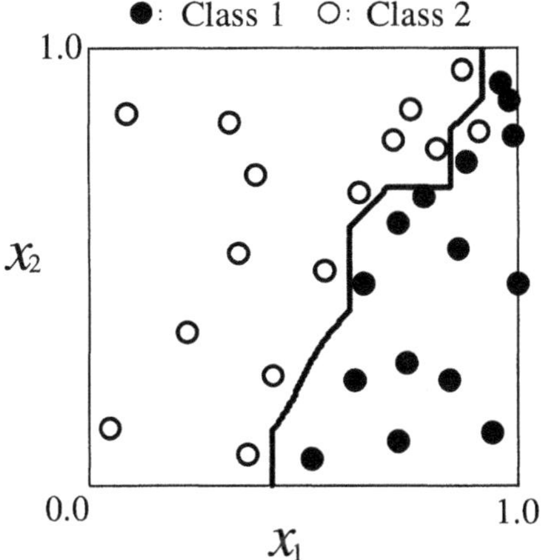

Fig. 4. Classification boundary

Now, we show the effect of the learning algorithm on the performance of the fuzzy rule-based classification system. We applied the learning algorithm of the grade of certainty in Subsection 2.4 to the fuzzy rule-based classification system in Table 1. The learning algorithm was iterated $N_{\text{learn}} = 100$ times for each given training pattern. The positive constants η_1 and η_2 were specified as $\eta_1 = 0.001$ and $\eta_2 = 0.01$. The classification boundary after the learning algorithm is shown in Fig. 5. From Fig. 5, we can see that all the given training patterns are correctly classified after the learning algorithm.

3 Fuzzy Rule-Based Ensembling System

In the last section, we explained fuzzy rule-based classification systems. That is, a fuzzy rule-based systems was used for determining the class of input

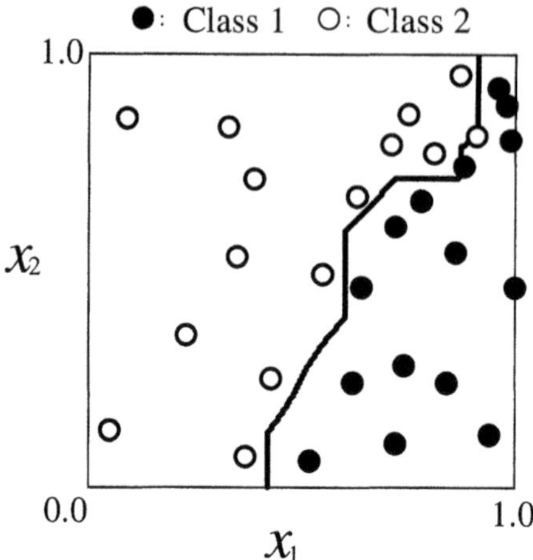

Fig. 5. Classification boundary after the learning of the grade of certainty

patterns. In this section, we introduce another type of the fuzzy rule-based systems that assign weight values to the classification results by the fuzzy rule-based classification systems. The weight can be viewed as a credit of the suggested classification results by a fuzzy rule-based classification system. The idea of the proposed method is shown in Fig. 1. The proposed method is composed of fuzzy rule-based classification systems described in Section 2, a single fuzzy rule-based ensembling system, and a gating node.

Let us assume that we have L fuzzy rule-based classification systems in our ensembling method. That is, each of the L fuzzy rule-based classification systems suggests its classification result for an input pattern. The fuzzy rule-based ensembling system estimates the weight value for each fuzzy rule-based classification system from the same input pattern. In the fuzzy rule-based ensembling system, fuzzy if-then rules of the following type are used:

$$\begin{aligned} &\text{Rule } R_k\text{: If } x_1 \text{ is } A_{k1} \text{ and } x_2 \text{ is } A_{k2} \text{ and } \ldots \text{and } x_n \text{ is } A_{kn} \\ &\qquad \text{then } \boldsymbol{w}_k = (w_{k1}, w_{k2}, \ldots, w_{kL}),\ k{=}1{,}2{,}\ldots{,}K, \end{aligned} \tag{13}$$

where $\boldsymbol{w}_k = (w_{k1}, w_{k2}, \ldots, w_{kL})$ is a consequent real-valued vector and K is the total number of fuzzy if-then rules in a fuzzy rule-based ensembling system. The fuzzy if-then rules in the fuzzy rule-based ensembling system is similar to those in the fuzzy classification rules in (1). The difference is that the consequent part is a real-valued vector in (13) while it is a class label with the grade of certainty in (1). The l-th element of the consequent vector of the k-th fuzzy if-then rule w_{kl} can be viewed as the degree of the credit for the classification result by the l-th fuzzy rule-based classification system.

3.1 Task of Fuzzy Rule-Based Ensembling System

Now, let us explain the task of our fuzzy rule-based ensembling system. Assume that we have m given training patterns $\boldsymbol{x}_p = (x_{p1}, x_{p2}, \ldots, x_{pn})$, $p = 1, 2, \ldots, m$. We also assume that L fuzzy rule-based classification systems have been already constructed by using the procedures in Section 2. From the given training patterns and the L fuzzy rule-based classification systems, we can examine which given training pattern is correctly classified by which fuzzy rule-based classification systems. Let $\boldsymbol{t}_p = (t_{p1}, t_{p2}, \ldots, t_{pL})$ be the target vector for the fuzzy rule-based ensembling system. The target vector $\boldsymbol{t}_p$ is specified from the corresponding training pattern $\boldsymbol{x}_p$ as follows:

$$t_{pl} = \begin{cases} 1, & \text{if the } l\text{-th fuzzy rule-based classification system} \\ & \text{correctly classifies the } p\text{-th training pattern } \boldsymbol{x}_p, \\ 0, & \text{otherwise.} \end{cases} \tag{14}$$

The main task of the fuzzy rule-based ensembling system is to determine the degree of the credit for each fuzzy rule-based classification system. That is, the fuzzy rule-based ensembling system rewards the fuzzy rule-based classification systems that suggest correct classification for an input pattern by assigning a high weight value. On the contrary, those fuzzy rule-based classification systems which suggest wrong classification should be penalized by assigning a low weight value. Thus our fuzzy rule-based ensembling system can be viewed as a pattern classification system that identifies the right fuzzy rule-based classification systems out of L systems. To summarize, the task of the fuzzy rule-based ensembling system is to assign appropriate weight values to the fuzzy rule-based classification systems in our ensembling method according to the given training patterns and the corresponding target vectors.

3.2 Constructing Fuzzy Rule-Based Ensembling System

In the last subsection, we explained the task of the fuzzy rule-based ensembling system. In this subsection, we describe how our fuzzy rule-based ensembling system is constructed from the given training data $\boldsymbol{x}_p$ and the target vectors $\boldsymbol{t}_p$, $p = 1, 2, \ldots m$.

The antecedent part of a fuzzy if-then rule can be specified as described in Subsection 2. That is, the antecedent fuzzy sets of a fuzzy if-then rule are specified by dividing each attribute of the pattern space $[0, 1]^n$. The consequent vector $\boldsymbol{w}_k$ is determined in a similar manner as the grade of certainty CF_j in Subsection 2.2:

[Determination of a weight vector $\boldsymbol{w}_k = (w_{k1}, \ldots, w_{kL})$]

1. Calculate $\beta_l(R_k)$ $(l = 1, \ldots, L)$ as

$$\beta_l(R_k) = \sum_{p=1 t_{pl}=1}^{m} \mu_{k1}(x_{p1}) \cdot \ldots \cdot \mu_{kn}(x_{pn}), \quad l = 1, 2, \ldots, L, \tag{15}$$

where μ_{kj}, $j = 1, 2, \ldots, n$ is the membership function of the antecedent fuzzy set A_{kj} and L is the number of fuzzy rule-based classification systems involved in our ensembling method.

2. The l-th element of the weight vector w_{kl} of the k-th fuzzy if-then rule R_k is determined as follows:

$$w_{kl} = \frac{\beta_l(R_k)}{\sum_{q=1}^{L} \beta_q(R_k)}, \quad l = 1, \ldots, L, \; k = 1, \ldots, K \tag{16}$$

Of course, the weight vector can be adjusted by a learning procedure. This will be explained in Section 4.6.

3.3 Calculating the Credit

In the last subsection, we explained how the fuzzy rule-based ensembling system can be constructed from the given training patterns and the corresponding target vector. In this subsection, we explain how the fuzzy rule-based ensembling system calculates the credit values for the classification results by the fuzzy rule-based classification results.

Assume that there are L fuzzy rule-based classification systems in our ensembling method. From an input vector $\boldsymbol{x} = (x_1, x_2, \ldots, x_n)$ each of the L fuzzy rule-based classification systems suggests its classification result for the input pattern. The fuzzy rule-based ensembling system offers additional information on how much such classification result from each fuzzy rule-based classification system is worth taking into consideration. The real-valued vector $\boldsymbol{W} = (W_1, W_2, \ldots, W_L)$ can be seen as the measure of the degree of the credit for the L fuzzy rule-based classification systems. We refer to this real-valued vector $\boldsymbol{W}$ as a credit vector. That is, the fuzzy rule-based ensembling system calculates $\boldsymbol{W}$ from an input vector $\boldsymbol{x}$. A credit vector for the classification results by fuzzy rule-based classification systems is calculated as follows:

[Fuzzy reasoning procedure for ensembling]

1. For each fuzzy if-then rule R_k, $k = 1, 2, \ldots, K$, calculate the compatibility of an input pattern $\boldsymbol{x}$ to the fuzzy if-then rule R_k as

$$\alpha_k(\boldsymbol{x}) = \mu_{k1}(x_1) \cdot \mu_{k2}(x_2) \cdot \ldots \cdot \mu_{kn}(x_n) \tag{17}$$

where $\mu_{ki}(\cdot)$, $i = 1, 2, \ldots, n$ is the membership function of the antecedent fuzzy set A_{ki}.

2. Determine the l-th element of the credit vector W_l as follows:

$$W_l = \frac{\sum_{k=1}^{K} \alpha_k(\boldsymbol{x}) \cdot w_{kl}}{\sum_{k=1}^{K} \alpha_k(\boldsymbol{x})}, \quad l = 1, 2, \ldots, L. \tag{18}$$

The suggested classes of an input vector by the fuzzy rule-based classification systems and the credit vector $\boldsymbol{W}$ from the fuzzy rule-based ensembling system is processed in the gating node (see Fig. 1). We explain the gating node in the next subsection.

3.4 Gating Node

The main task of the gating node is to determine the final classification for an input vector by using the suggested classes from the fuzzy rule-based classification systems and the credit vector $\boldsymbol{W}$ calculated by the fuzzy rule-based ensembling system. There are various ideas for finally determining the classification of an input pattern. In this chapter, we examine two decision schemes for the final classification. One is a single winner scheme of the weight value, and the other is a single winner scheme of the product.

In the single winner scheme of the weight value, the final classification is dependent only on the classification result by the single fuzzy rule-based classification system that has the largest weight value W_l. That is, the gating node examine the weight value W_l, $l = 1, \ldots, L$ in the weight vector $\boldsymbol{W}$ from the fuzzy rule-based ensembling system to adopt the classification result suggested by the most creditable fuzzy rule-based classification system.

In the single winner scheme of the product, the gating node takes both the degree of the credit (that is, the credit vector $\boldsymbol{W}$) and the certainty level of the classification of the fuzzy rule-based classification system into consideration. We measure the certainty level of the classification by $\alpha_{\text{Class } h_p^*}(\boldsymbol{x})$ in (8). The final classification is the classification result suggested by the single winner fuzzy rule-based classification system that has the maximum product of the certainty level (i.e., $\alpha_{\text{Class } h_p^*}(\boldsymbol{x})$) and the degree of the credit (i.e., W_l).

We have also examined other types of decision schemes such as weighted majority vote. However, their performance was not good in our preliminary experiments. Thus we will show only the results by two single winner schemes.

4 Computer Simulations

In this section, we examine the performance of our proposed ensembling method. First we explain three real-world data sets which are used in our computer simulations to examine the performance of our ensembling method.

Next, we explain how the ensembling method is designed for our experiments. Then we compare the performance of our ensembling method to other pattern classification systems in literature.

4.1 Test Problems

In order to examine the performance of the ensembling method, we use three real-world pattern classification problems: iris data set, appendicitis data set, and cancer data set.

The iris data set is a four-dimensional three-class problem with 150 given training patterns [24]. There are 50 training patterns from each class. This data set is one of the most well-known pattern classification problems. Many researchers have applied their classification methods to the iris data set. For example, Weiss and Kulikowski [25] examined the performance of various classification methods such as neural networks and nearest neighbor classifier for this data set. Grabisch and Dispot [26] has also examined the performance of various fuzzy classification methods such as fuzzy integrals and fuzzy k-nearest neighbor for the iris data set.

Weiss and Kulikowski [25] has also examined the performance of various classification methods for appendicitis and cancer data sets. The appendicitis data set consists of 106 given training patterns with seven attributes. The cancer data set is a nine-dimensional two-class pattern classification problem. In Grabisch's works [26–28], various fuzzy classification methods have been applied to appendicitis and cancer data sets in order to compare each of those fuzzy classification methods.

4.2 Experimental Settings

As is already described, our ensembling method consists of a single fuzzy rule-based ensembling system, several fuzzy rule-based classification systems, and a gating node. In our computer simulations in this chapter, we use four fuzzy rule-based classification systems in the proposed ensembling method. In the design of the fuzzy rule-based classification systems, we vary the number of fuzzy sets in each attribute. That is, we specify the number of fuzzy sets for each attribute as two, three, four, and five. For example, when we apply our ensembling method to the iris data set (four-dimensional problem), the total number of fuzzy if-then rules in the four fuzzy rule-based classification systems is $2^4 = 16$, $3^4 = 81$, $4^4 = 256$, and $5^4 = 625$. The membership functions when each attribute is divided into two, three, four, and five fuzzy sets are shown in Fig. 2 (a), (b), (c), and (d), respectively.

We also examine the effect of selecting a small number of attributes on the performance of the proposed ensembling method. When we construct a fuzzy rule-based classification system, only the selected attributes are allowed to be included in the fuzzy if-then rules. We decide whether an attribute is used or not based on the entropy measure [29–31]. First we calculate the entropy value $E(S,T)$ of each attribute according to the fuzzy subset as follows:

$$E(S,T) = \frac{|S_1|}{|S|}\sum_{k=1}^{c}\frac{|S_{1k}|}{|S_1|}\log_2\frac{|S_1|}{|S_{1k}|} + \frac{|S_2|}{|S|}\sum_{k=1}^{c}\frac{|S_{2k}|}{|S_2|}\log_2\frac{|S_2|}{|S_{2k}|}, \quad (19)$$

where S is the training data set, T is a threshold value, S_1 and S_2 is subsets of training data devided by the threshold value T, S_{ij} is the set of training pattern from Class j in the subset S_i, and $|\cdot|$ is the cardinality of a data set. The candidate thresholds are determined from the shape of the membership functions. That is, we specify the threshold values as the crossing points of adjacent menbership functions. Then a small number of most separable attribute is selected by choosing the attributes that have small entropy measure.

When we construct the fuzzy rule-based ensembling system, each attribute is divided into three fuzzy sets (see Fig. 2 (b)). Also we restrict the number of antecedent fuzzy sets up to two so that we can avoid the explosive increase in the total number of generated fuzzy if-then rules. Thus, the total number of fuzzy if-then rules in the fuzzy rule-based ensembling system (that is, K) is $K = {}_nC_0 + {}_nC_1 \cdot 3 + {}_nC_2 \cdot 3^2$ where n is the number of attributes in a pattern classification problem at hand.

For the gating node, we examine both the single winner scheme of the weight value and the single winner scheme of the prodoct.

4.3 Performance Evaluation on Iris data

From this subsection, we show the performance of our proposed ensembling method for each pattern classification problems on given training data. For the iris data set, we first construct four fuzzy rule-based classification systems by dividing each selected attribute into two, three, four, and five fuzzy sets. Then the fuzzy rule-based ensembling system is designed from the given training patterns and the target vectors (see Section 3). Also we examined the effect of selecting the number of attributes on the performance. Classification results are shown in Table 2. In Table 2, we show the number of the selected attributes, the performance of the best single fuzzy rule-based classification system in our ensembling method, the performance of the single winner scheme of the weight, and the performance of the single winner scheme of the product.

From Table 2, we can see that our ensembling method is exactly the same as the best single fuzzy rule-based classfication system. This is because the iris data set is not a challenging problem with small overlaps between different classes. In that case, the ensembling method may not work. This problem will be explained again for the performance on unseen test data in Subsection 4.7.

Table 2. Classification result on training data (Iris data set)

# of attributes	Best single	Winner of weight	Winner of product	# of rules
2	95.3%	95.3%	95.3%	54
3	96.0%	96.0%	96.0%	224
4	96.0%	96.0%	96.0%	978

We can also see that the total number of fuzzy if-then rules which are involved in the ensembling method is more than the number of training patterns when the number of attributes is more than two. This problem is avoided by applying genetic algorithms for constructing compact fuzzy rule-based classification systems. For example, we can generate compact fuzzy rule-based classification systems with less than ten fuzzy if-then rules. This issue will be mentioned in Subsection 4.7.

4.4 Performance Evaluation on Appendicitis data

In the same manner as for the iris data set in the last subsection, we examined the performance of the proposed ensembling method for the appendicitis data set. For the appendicitis data set, we only generated fuzzy if-then rules with two antecdent fuzzy sets in constructing fuzzy rule-based classification systems. This is because the number of possible fuzzy if-then rules is extremely huge when we divide all the seven attributes (for example, $5^7 = 78125$ fuzzy if-then rules are generated when each attribute is divided into five fuzzy sets). The performance of our ensembling method on appendicitis data set is shown in Table 3.

Table 3. Classification result on training data (Appendicitis data set)

# of attributes	Best single	Winner of weight	Winner of product	# of rules
2	89.6%	89.6%	89.6%	54
3	87.7%	87.7%	87.7%	224
4	87.7%	88.7%	88.7%	978
5	89.6%	89.6%	89.6%	4404
6	87.7%	88.7%	88.7%	20424
7	88.7%	88.7%	88.7%	96324

4.5 Performance Evaluation on Cancer data

We evaluate the performance of the ensembling method in almost the same manner as in the last section. That is, we restrict the number of the antecedent fuzzy sets to two. The classification results is shown in Table 4.

From Table 4, we can see that both ensembling schemes (the single winner scheme of weight values and the single winner scheme of product) slightly

Table 4. Classification result on training data (Cancer data set)

# of attributes	Best single	Winner of weight	Winner of product	# of rules
2	76.6%	77.3%	77.3%	54
3	77.3%	77.3%	77.3%	224
4	77.3%	77.3%	77.3%	978
5	75.5%	76.9%	76.9%	4404
6	76.2%	76.9%	76.9%	20424
7	76.2%	76.9%	76.9%	96324
8	75.9%	76.2%	76.2%	460478
9	75.9%	76.2%	76.2%	2222964

outperformed the best single fuzzy rule-based classification system. However, the difference between the best single fuzzy rule-based classification system and our proposed ensemble is not large. In the next subsection, we try to improve the ensemble method by adjusting the consequent real-vector.

4.6 Learning of Consequent Weight Vectors

In this subsection, we introduce the learning schemes of the weight vector from the fuzzy rule-based ensembling system. Let us assume that there are k fuzzy if-then rules in the fuzzy rule-based ensembling system. We also assume that the output vector can be denoted as $\boldsymbol{W} = (W_1, W_2, \ldots, W_L)$ where L is the number of fuzzy rule-based classification systems. After the final classification of an input vector $\boldsymbol{x}$, the consequent real-valued vector $\boldsymbol{w}_k = (w_{k1}, w_{k2}, \ldots, w_{kL})$ is adjusted as follows:

$$w_{kl}^{\text{new}} = \begin{cases} w_{kl}^{\text{old}} + \eta_1 \cdot \dfrac{\alpha_k(\boldsymbol{x})}{\sum_{q=1}^{K} \alpha_q(\boldsymbol{x})} \cdot (1 - w_{kl}^{\text{old}}), & \text{if the class suggested by } l\text{-th fuzzy system is correct,} \\[2ex] w_{kl}^{\text{old}} - \eta_2 \cdot \dfrac{\alpha_k(\boldsymbol{x})}{\sum_{q=1}^{K} \alpha_q(\boldsymbol{x})} \cdot w_{kl}^{\text{old}}, & \text{if the class suggested by } l\text{-th fuzzy system is not correct,} \end{cases} \tag{20}$$

where η_1 and η_2 are positive constants and α_k is calculated in (17). This learning algorithm is a modified version of the learning algorithm of the grade of certainty of the fuzzy if-then rules in fuzzy rule-based classification systems in (10) and (11).

Let us demonstrate the ability of the learning algorithm of the consequent vectors. For the appendicitis data set, we applied the learning algorithm to the fuzzy rule-based ensembling system in Subsection 4.4. This learning algorithm is iterated for 5000 times for each training pattern. We show the classification results in Table 5 where we used the single winner scheme of the weight value for deciding the final classification of an input pattern.

Table 5. Classification results after the learning of w_{kl} (Appendicitis data set)

	# of learning iterations				
# of attributes	0	100	1000	2000	5000
2	89.6%	90.6%	92.5%	92.5%	91.5%
3	87.7%	87.7%	90.6%	90.6%	90.6%
4	88.7%	89.6%	91.5%	91.5%	89.6%
5	89.6%	91.5%	89.6%	89.6%	90.6%
6	88.7%	87.7%	90.6%	90.6%	90.6%
7	88.7%	88.7%	89.6%	89.6%	90.6%

From Table 5, we can see that the performance of the proposed ensembling method is improved by learning the consequent weight vector $\boldsymbol{w}_k$, $k = 1, 2, \ldots, K$.

4.7 Generalization ability of the ensembling method

It is often said that the most important issue in designing classification system is how well that classification system generalizes. That is, classification systems should correctly classify unseen input patterns. To examine the generalization ability of the proposed ensembling method, we applied the leaving-one-out procedure for the iris and the appendicitis data sets, and 10-fold cross validation method for the cancer data set. We iterated the procedure of the 10-fold cross validation method for ten times for the cancer data set. The simulation results are shown in Table 6 to Table 8. Since we obtained similar results between the single winner scheme of weight value and the single winner scheme of product, we only show the results by the single winner scheme of weight values. Ishibuchi et al.[20] used a GA-based algorithm for constructing a fuzzy rule-based classification system. The performance of the fuzzy rule-based classification system was 94.67% on unseen data after 1000 generations with a population of 10 individuals. That is, they discovered the fuzzy rule-based classification system after searching a total of 10000 fuzzy rule-based classification systems. On the other hand, we didn't find good fuzzy rule-based classification systems. The fuzzy rule-based classification systems were generated in a heuristic manner. However, the advantage of the Ishibuchi et al.'s method is the number of fuzzy rules involved in a fuzzy rule-based systems. Their fuzzy rule-based classification systems consist of around 20 fuzzy if-then rules. It is expected that we can generate good fuzzy rule-based classification systems with a small number of fuzzy rules by using the genetic algorithm. We implemented the GA-based algorithm as in [20] for generating a fuzzy rule-based classification system. That is, we executed the GA-based classification system in the same manner as in Ishibuchi et al.[20] with different initial populations four times to obtain four different fuzzy rule-based classification system. The best individual after the execution of the genetic algorithm was employed as a member of our ensembling system. The result of such GA-based classification is as follows:

Average Classification rate: 97.3%
Average total number of fuzzy rules : 22.7

From the result, we can see that the number of rules were decreased by the GA-based method. The average number of fuzzy if-then rules per a single fuzzy rule-based classification system is $22.7/4 \approx 5.7$. Considering that 978 rules were involved at the maximum in the ensembling method in Table 2, we could reduced the number of fuzzy if-then rules into less than 3% by using the genetic algorithms. We don't go into this issue in this chapter since it is beyond the scope of this chapter. It will be treated in our future research.

Table 6. Classification results on unseen data (Iris data set)

	# of learning iterations			
# of attributes	0	50	100	150
2	95.3%	92.7%	94.7%	94.7%
3	95.3%	92.7%	95.3%	94.0%
4	95.3%	92.7%	94.0%	94.0%

Table 7. Classification results on unseen data (Appendicitis data set)

	# of learning iterations			
# of attributes	0	50	100	150
2	87.7%	85.8%	85.8%	84.0%
3	85.8%	83.0%	83.0%	84.0%
4	86.8%	85.8%	84.9%	84.9%
5	82.1%	83.0%	84.0%	84.0%
6	83.0%	83.0%	83.0%	84.0%
7	85.8%	84.9%	84.9%	84.9%

To compare the performance of our ensembling method to other classification methods, Table 9 summarizes the classication results reported in [25] and the results by the fuzzy classification methods reported in [26–28].

Comparing these results with the results by our ensembling method, we can generally say that the performance of our ensembling method is better than or comparable to the performance of the other classification results. For the iris data set, the performance of the ensembling method is not good. On the other hand, for the cancer data set the ensembling method performs very well.

Table 8. Classification results on unseen data (Cancer data set)

	# of learning iterations			
# of attributes	0	50	100	150
2	74.1%	75.1%	75.8%	76.0%
3	75.2%	74.8%	74.5%	74.5%
4	75.0%	74.5%	73.8%	73.6%
5	74.3%	73.9%	73.5%	73.3%
6	74.5%	74.4%	74.2%	74.1%
7	74.5%	74.5%	74.3%	74.5%
8	73.7%	73.7%	73.7%	73.7%
9	72.6%	72.6%	72.6%	72.7%

Table 9. Classification results reported in [25–28].

		Best	Average	Worst
Iris	[25]	98.0%	95.2%	84.0%
	[26–28]	97.7%	94.4%	91.3%
Appendicitis	[25]	89.6%	84.3%	73.6%
	[26–28]	86.8%	81.7%	71.2%
Cancer	[25]	77.1%	70.9%	65.6%
	[26–28]	68.0%	60.6%	54.9%

5 Conclusions

In this chapter, we proposed an ensembling method for pattern classification problems. In our ensembling method, two different types of fuzzy rule-based systems were used. One is a fuzzy rule-based classification system that suggests the class of an input vector. The other type of the fuzzy rule-based system is a fuzzy rule-based ensembling system that determines which classification systems should be used for the classification. The proposed ensembling method consists of one fuzzy rule-based ensembling system, several fuzzy rule-based classification systems, and a gating node that is used for final classification. As a gating node, we consider the single winner scheme for weight values and the single winner scheme for the product. In computer simulations, we examined the performance of our ensemble learning method on three real-world pattern classification problems. Simulation results showed that the performance of our ensembling method is better than the best single fuzzy rule-based classification system for given training patterns. We also showed that the performance of our method on unseen patterns was better than the other classification results. However, the performance of the ensembling method is somewhat poor for iris data set where there is no large overlaps between different classes. The reason for that is remained for our future research.

The major problem with the ensembling method is that the total number of fuzzy if-then rules involved in the ensembling method can be intractably large. In order to avoid this problem, we gave a suggestion of applying a genetic-algorithm-based rule selection method to the construction of compact fuzzy rule-based classification systems.

The other future work includes how to constitute the member of the fuzzy rule-based classification systems. It is possible that each fuzzy rule-based classification system is generated so that only those patterns from one class can be correctly classified. Then collecting such a fuzzy rule-based classifcation system from each class, we could design a more powerful ensembling method. The design of the fuzzy rule-based ensembling method automatically by using some machine learning technique such as neural networks is also another research direction.

References

1. Sugeno M. (1985) An Introductory Survey of Fuzzy Control. Information Science Vol. 30, No. 1/2:59–83
2. Lee C. C. (1990) Fuzzy Logic in Control Systems: Fuzzy Logic Controller Part I and Part II. IEEE Trans. Syst., Man, Cybern. Vol. 20:404–435
3. Leondes C. T. (Ed.) (1999) Fuzzy theory Systems: Techniques and Applications. Academic Press, San Diego, Vol. 1–4
4. Ishibuchi H., Nakashima T. (1999) Performance evaluation of fuzzy classifier systems for multi-dimensional pattern classification problems. IEEE Trans. on Syst., Man, Cybern. Part B Vol. 29:601–618
5. Ishibuchi H., Nakashima T. (1999) Improving the performance of fuzzy classifier systems for pattern classification problems with continuous attributes. IEEE Trans. on Industrial Electronics Vol. 46, No. 6:1057–1068
6. Yuan Y., Zhang H. (1996) A genetic algorithms for generating fuzzy classification rules. Fuzzy Sets and Systems, Vol 84, No. 1:1–19
7. Ueda N., Nakano R. (1996) Generalization error of ensemble estimators. Proc. of Intl. Conf. on Neural Networks, 90–94
8. Battiti, R. and Colla, A. M. (1994) Democracy in neural nets: Voting schemes for classification. Neural Networks Vol. 7:691–707
9. Xu L., Krzyzak A., Suen C. Y. (1992) Methods of combining multiple classifiers and their applications to handwriting recognition. IEEE Trans. on Syst., Man, Cybern. Vol. 22, No. 3:418–435
10. Benediktsson J. A., Colla A. M. (1992) Consensus theoretic classifcation methods. IEEE Trans. on Syst., Man, Cybern. Vol. 22, No. 4:688–704
11. Wolpert D. H. (1992) Stacked generalization. Neural Networks, Vol. 5:241–259
12. Jacobs R. A., Jordan M. I. (1991) Adaptive mixtures of local experts. Neural Computation, Vol. 3:79–87
13. Hansen L. K., Salamon P. (1990) Neural network ensembles. IEEE Trans. on Pattern Analysis and Machine Intelligence, Vol. 12, No. 10:993–1001
14. Cho S. B. (1995) Fuzzy aggregation of modular neural networks with ordered weighted averaging operators. Intl. J. Approx. Reasoning Vol.=13:359–375
15. Cho S. B., Kim J. H. (1995) Combining multiple neural networks by fuzzy integral for robust classification. IEEE Trans. on Syst., Man, Cybern. Vol. 25, No. 2:380–384

16. Cho S. B., Kim J. H. (1995) Multiple network fusion using fuzzy logic. IEEE Trans. on Neural Networks Vol. 6, No. 2:497–501
17. Ishibuchi H., Nakashima T., Morisawa T. (1999) Voting in fuzzy rule-based systems for pattern classification problems. Fuzzy Sets and Systems, Vol. 103, No. 2:223–238
18. Klir G.J., Yuan B., Fuzzy Sets and Fuzzy Logic. Prentice-Hall, 1995.
19. Ishibuchi H., Nozaki K., Tanaka H. (1992) Distributed representation of fuzzy rules and its application to pattern classification. Fuzzy Sets and Systems, Vol. 52, No. 1:21–32
20. Ishibuchi H., Nozaki K., Yamamoto N., Tanaka H. (1995) Selecting fuzzy if-then rules for classification problems using genetic algorithms. IEEE Trans. on Fuzzy Systems. Vol. 3, No. 3:260–270
21. Ishibuchi H., Nakashima T. (2001) Effect of rule weights in fuzzy rule-based classification systems. IEEE Trans. on Fuzzy Systems. Vol. 9, No. 4:506–515
22. Nozaki K., Ishibuchi H., Tanaka H. (1996) Adaptive fuzzy rule-based classification systems. IEEE Trans. on Fuzzy Systems, Vol. 4, No. 3:238–250
23. Ishibuchi H., Nakashima T., Morisawa T. (1999) Voting in fuzzy rule-based systems for pattern classification problems. Fuzzy Sets and Systems, Vol. 103, No. 2:223–238
24. Fisher R. A. (1936) The use of multiple measurements in taxonomic problems. Ann. of Eugenics, Vol. 7:179–188
25. Weiss S. M., Kulikowski C. A. (1991) Computer systems that learn. Morgan Kaufmann, San Mateo
26. Grabisch M., Dispot F. (1992) A comparison of some methods of fuzzy classification on real data. Proc. of 2nd Intl. Conf. on Fuzzy Logic and Neural Networks, 659–662
27. Grabisch M. (1996) The representation of importance and interaction of features by fuzzy measures. Pattern Recognition Lett, Vol. 17:567–575
28. Grabisch M., and Nicolas J. -M. (1994) Classification by fuzzy integral: performance and tests. Fuzzy Sets and Systems Vol. 65, No. 2/3:255–271
29. Quinlan J. R. (1993) C4.5: Programs for machine learning. Morgan Kaufmann Publishers, San Mateo
30. Dougherty J., Kohavi R., and Sahami M. (1995) Supervised and unsupervised discretization of continuous features. Proc. of the 12th Intl. Conf. of Machine Learning, 194–202
31. Nakashima N., and Ishibuchi H. (2001) Supervised and unsupervised fuzzy discretization of continuous attributes for pattern classification problems. Proc. of Knowledge-Based Intelligent Information Engineering Systems & Allied Technologies, 32–36

Part 4:

Information Extraction

Data Mining Using Granular Linguistic Summaries

Ronald R. Yager

Machine Intelligence Institute, Iona College, New Rochelle, NY 10801

Abstract. The concept of linguistic summaries is introduced as a user friendly way to present information mined from a database. We discuss methods for measuring the amount of information provided by a linguistic summary. The issue of conjecturing, how to decide on which summaries maybe informative, is discussed. Two approaches to help us focus on relevant summaries are discussed. The first, called the template method, makes use of available linguistic concepts related to the domain of the attributes involved. The second approach uses the mountain clustering method to help focus our summaries.

1 Introduction

Database mining is concerned with the task of discovering knowledge implicit in large databases [1–11]. Wide spread interest exists in this technology in no small part do to the vast amounts of information being made available as a result of confluence of communications and computing provided by the internet [12,13]. A fundamental feature of data mining, one that distinguishes it from database querying, is that the knowledge discovered is based upon data contained in multiple records rather then in a single record. Thus whereas querying is based upon the process of retrieval, essentially a matching or comparing operation, data mining generally involves a step of fusing data from multiple records. Thus technologies related to fusion and aggregation are central to the development of data mining techniques.

An issue of central importance to the task of data mining is the matter of the representation of the knowledge discovered. In [14] we originally suggested the use of linguistic summaries as a tool for representing various kinds of knowledge discovered in a database. In [15–24] this idea has been applied and extended. The key idea behind these summaries is to use linguistic terms, words from natural language, in presenting the information discovered. This approach is very much in the spirit of what Zadeh has called computing with words [25,26]. Implicit in this is the realization that human cognition is most effectively suited to manipulate the types of granular of concepts conveyed by natural language rather then precise and brittle concepts.

An important problem in database mining is the issue of conjecturing and focusing, deciding where to look for meaningful summaries. In order to deal with this problem we generally can benefit from some knowledge of the domain in which we are mining. Two methods are discussed for addressing this issue, a template approach and a clustering approach. In the template based method we suggest using a predetermined template vocabulary to provide a basic set of terms used in the summaries. By providing these templates we are essentially using our-knowledge of the vocabulary that is commonly used in discussing an attribute of interest. One considerable advantage of this template approach is that the results of our data mining can easily integrated and used with other existing knowledge that may be available. The clustering method, which uses the mountain method clustering algorithm [27–29], can be seen as using the information in the database itself as providing some direction for focusing. One advantage of this approach is that it is more open to uncovering knowledge that requires a non standard view of a domain.

2 Linguistic Summaries

Assume V is some attribute in a database having as its domain the set X. We allow V to assume either numeric or non-numeric values. Associated with V is a bag D consisting of the values for V assumed by the objects in the database, $D = [a_1, a_2, \ldots, a_n]$. A linguistic summary associated with V is a global statement based upon the value in D. If V is the attribute age some examples of simple linguistic summaries are:

Most people in the database are about 25 years old
Few people in the database are old
Nearly a quarter of the people in the database are middle aged

Formally a simple linguistic summary is a statement of the form:

Q *objects in the database have* V *is* **S**.

In the above **S** is called the summarizer and **Q** is called the quantity in agreement. Also associated with a linguistic summary is a measure of validity of the summary, τ. The value τ is used to indicate the truth of the statement that **Q** objects have the property that V is **S** in the light of the data set D.

A fundamental characteristic of this formulation is that the summarizer and quantity in agreement are expressed in linguistic terms. One advantage of using these linguistic summaries is that we can provide statements about the database in terms that are very easy for people to comprehend. A second advantage, one that will be useful in data mining, is that these statements can allow concepts having large granularity.

Using fuzzy subsets we are able to provide a formal semantics for the terms used in a linguistic summary. In a procedure to be subsequently described, we shall use this ability to formalize the summarizers and quantity in agreement to evaluate the validity of the linguistic summary. This validation process will be based upon a determination of the compatibility of the linguistic summary with the data set D. It should be pointed out that for a given attribute we can conjecture numerous different summaries, then with the aid of the data set D we can evaluate τ to determine which are the valid summaries.

In developing our approach to validating a linguistic summary considerable use will be made of the ability to represent a linguistic summarizer by a fuzzy subset over the domain of the attribute. If V is some attribute taking its value from the domain X and if **S** is some concept associated with this attribute we can represent **S** by a fuzzy subset S on X such that for each $x \in X$, $S(x) \in [0, 1]$ is the degree of compatibility of the value x with the concept **S**. If V is age and **S** is the concept middle age then $S(40)$ indicates the degree to which 40 years old is compatible with the idea of middle age. Even in environments in which the underlying domain is non-numeric using this approach allows us to obtain numeric values for the membership grade in the fuzzy subset. For example if V is the attribute city of residence which takes as its domain the cities in the U.S. we can express the concept "lives near New York" as a fuzzy subset. The second component in our linguistic summary is the quantity in agreement **Q**. These objects belong to a class of concepts called linguistic quantifiers [30] . Examples of linguistic quantifiers are terms such as most, few, about half, all. Essentially linguistic quantifiers are fuzzy proportions. An alternative view of these objects are as generalized logical quantifiers [31]. In [30] Zadeh suggested we can represent these linguistic quantifiers as fuzzy subsets of the unit interval. Using this representation the membership grade of any proportion $r \in [0, 1]$, $Q(r)$, is a measure of the compatibility of the proportion r with the linguistic quantifier we are representing by the fuzzy subset **Q**. For example if **Q** is the quantifier MOST then $Q(0.9)$ represents the degree to which 0.9 satisfies the concept most.

In [32,33] Yager identified three classes of linguistic quantifiers that cover most of those used in natural language. (1) Q is said to be monotonically non-decreasing if $r_1 > r_2 \rightarrow Q(r_1) \geq Q(r_2)$. Examples of this type of quantifiers are at least 30%, most, all. (2) A quantifier Q is said to be monotonically non-increasing if $r_1 > r_2 \rightarrow Q(r_1) \leq Q(r_2)$. Examples of this type of quantifiers are at most 30%, few, none. (3) A quantifier Q is said to be unimodal if there exists two values $a \leq b$ both contained in the unit interval such that for $r < a$, Q is monotonically non-decreasing, for $r > b$, Q is monotonically non-increasing and for $r \in [a, b]$, $Q(r) = 1$. An example of this type of quantifier is "about 0.3".

An important idea that can be associated with a linguistic quantifier is the concept of an antonym. If Q is a linguistic quantifier its antonym is also a linguistic quantifier, denoted $\hat{Q}$, such that $\hat{Q}(r) = Q(1 - r)$. The operation of taking an antonym is involutionary, that is $\hat{\hat{Q}} = Q$. From this we see that antonyms come in pairs. Prototypical examples of antonym pairs are **all-none** and **few-many**. Consider the quantifier ***at most 0.3*** defined as $Q(r) = 1$ if $r \leq 0.3$ and $Q(r) = 0$ if $r > 0.3$. Its antonym has $\hat{Q}(1 - r) = 1$ if $r \leq 0.3$ and $\hat{Q}(1 - r) = 0$ if $r \geq 0.3$. This can be seen to be equivalent to

$\hat{Q}(r) = 1$ if $r \geq 0.7$ and $\hat{Q}(r) = 0$ if $r < 0.7$. Thus the antonym of ***at most 0.3*** is ***at least 0.7***.

Care must be taken to distinguish between the antonym of a quantifier and its negation. We recall the negation of Q denoted $\bar{Q}$ is defined such that $\bar{Q}(r) = 1 - Q(r)$. We see that the negation of ***at most 0.3*** is $\bar{Q}(r) = 0$ if $r \leq 0.3$ and $\bar{Q}(r) = 1$ if $r \geq 0.3$ which corresponds ***"to at least 0.3"***.

Figure 1 graphically illustrates these distinctions.

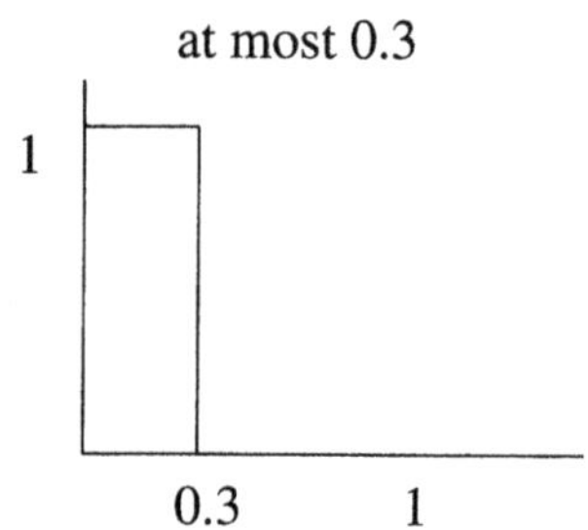

Fig. 1. Distinction between antonym and negation

Having discussed the concepts of summarizer and quantity in agreement we are now in a position to describe the methodology used to calculate the validity τ of a linguistic summary. Assume $D = [a_1, a_2, \ldots, a_n]$ is the collection of values that appear in the database for the attribute V. Consider the linguistic summary:

Q items in the database have values for V that are S.

The basic procedure to obtain the validity τ of this summary in the face of the data D is: [14]

(1) For each $a_i \in D$, calculate $S(a_i)$, the degree to which a_i satisfies the summarizer S.
(2) Let $r = \frac{1}{n}\sum_{i=1}^{n} S(a_i)$, the proportion of D that satisfy S.

(3) $\tau = Q(r)$, the grade of membership of r in the proposed quantity in agreement.

Examples: Assume we have a database consisting of 10 entries. Let D be the collection of ages associated with these entries: $D =< 30, 25, 47, 33, 29, 50, 28, 52, 19, 21 >$

1. Consider the linguistic summary *most people are "at least 25"*. In this case the summarizer "at least 25" can be expressed simply as $S(x) = 0$ if $x < 25$ and $S(x) = 1$ if $x \geq 25$. We define the quantity in agreement "most" by the fuzzy subset $Q(r) = 0$ if $r < 0.5$ and $Q(r) = (2r - 1)^{1/2}$ if $r \geq 0.5$. For the first eight items in the data collection $S(x) = 1$ while for the remaining two items $S(x) = 0$. Thus in this case $r = 8/10$ and hence $\tau = Q(8/10) = 0.77$
2. Consider the linguistic summary *"about half the ages are near 30"*. Here we define "near thirty" by $S(x) = exp^{-(\frac{x-30}{25})^2}$ and we define "about half" as $Q(r) = exp^{-(\frac{r-0.5}{0.25})^2}$ In this case $r = 3.94/10 = 0.394$ and $Q(0.394) = 0.95$.
3. Consider the proposition *"most of the people are young"*. We define young as $S(x) = 1$ if $x < 20$, $S(x) = -\frac{1}{10}x + 3$ if $20 \leq x \leq 30$ and $S(x) = 0$ if $x > 30$. In this case $r = 0.27$ and using our previous definition of most we get $\tau = Q(0.27) = 0$.

A number of interesting properties can be associated with these linguistic summaries [34]. Consider a proposed summary "**Q** items have V is **S**" and assume data set D has cardinality n. The associated validity is $\tau = Q(\frac{\sum S(x)}{n})$. Now consider the summary

$\hat{Q}$ items have V is not S

where $\hat{Q}$ is the antonym of Q. In this case the measure of validity τ' is

$$\tau' = \hat{Q}(\frac{\sum \bar{S}(x)}{n}) = \hat{Q}(\frac{\sum 1 - S(x)}{n}) = \hat{Q}(\frac{n - \sum S(x)}{n})$$

however since $\hat{Q}(1 - r) = Q(r)$ we see that $\tau' = \tau$. The two linguistic summaries have the same measure of validity. The prototypical manifestation of this is that the summary ***Most** people are young* will have the same validity as ***Few** people are **not** young.*

Consider now the summary "**Not** Q objects are S." In this case

$$\bar{Q}(\frac{\sum S(x_i)}{n}) = 1 - Q(\frac{\sum S(x_i)}{x}) = 1 - \tau$$

This statement has a validity complement to our original proposition. From this we see that

$$\tau(S, Q) + \tau(S, \bar{Q}) = 1.$$

Thus far we have considered linguistic summaries involving only one attribute. The approach described above can be extended to the case of multiple attributes from a database. We shall first consider summaries of this form

Most people in the database are tall ***and*** *young.*

Assume U and V are two attributes appearing in the database. Let R and S be concepts associated with each of these attributes respectively the generic form of the above linguistic summary is

"Q people in the database have U is R and V is S."

In this case our data consists of a collection of pairs, $D =< (a_1, b_1), (a_2, b_2), \ldots, (a_n, b_n) >$ where a_i is a value for attribute V and b_i is a value for attribute U. Our procedure for obtaining the validity of the linguistic summary in this case is

1. For each i calculate $R(a_i)$ and $S(b_i)$.
2. Let $r = \frac{1}{n} \sum_{i=1}^{n} (R(a_i) S(a_i))$
3. $\tau = Q(r)$

We should note that in step two we can replace the product of $R(a_i)$ and $S(a_i)$ by any t-norm such as $\text{Min}(R(a_i), S(a_i))$. However we shall use the product. The above procedure can be easily generalized to consider any number of attributes.

We also can consider linguistic summaries of the forms

Q objects have V is S **or** U is R.

In this case the procedure is the same except in step two the product is replaced by a union operation, t-conorm, such as $\text{Max}(R(a_i), S(b_i))$ or $R(a_i) + S(a_i) - R(a_i)S(a_i)$.

We now consider another class of linguistic summaries manifested by statements like

Most <u>tall</u> people in the database are <u>young</u>.

This form of linguistic summary is related to the type of association rule discovery that is of great interest in many applications of data mining [9]. The linguistic summary expressed here can be equivalently expressed as

In ***most*** cases of our data: if a person is **tall** then they are **young**

Here then we have an embedded association rule

if height is **tall** then age is **young**

Furthermore we are qualifying our statement of this association rule with the quantifier *most.*

In this case we have as our generic form

Q of the U is **R** objects in the database have V is **S**

In the above we call **R** the **qualifier** of the summary. The procedure for calculating the validity of this type of linguistic summary has the same three step process

1. For each i calculate $R(a_i)$ and $S(b_i)$.
2. $r = \frac{\sum_{i=1}^{n} R(a_i) S(b_i)}{\sum_{i=1}^{n} R(a_i)}$

3. $\tau = Q(r)$

A fundamental distinction between this and the previous cases is in step two, here instead of dividing by n, the number of objects in the data base, we divide by the number of objects having R.

We note that we can naturally extend this procedure to handle summaries of the form

Most young people are tall and live near New York
Few well paid and young people in the database live in the suburbs.

Consider the linguistic summary corresponding to the quantified association rule

$$QR \text{ are } S$$

and the related summary "$\hat{Q}R$ are $\bar{S}$" where $\hat{Q}$ is the antonym of Q and $\bar{S}$ is the negation of S. The validity of the first summary is $\tau_1 = Q(r)$ where $r = \frac{\sum_{i=1}^{n} R(a_i)*S(b_i)}{\sum_{i=1}^{n} R(a_i)}$. The validity of the second summary is $\tau_2 = \hat{Q}(r^*)$ where $r^* = \frac{\sum_{i=1}^{n} R(a_i)*(1-S(b_i))}{\sum_{i=1}^{n} R(a_i)} = 1 - r$. However we note

$$\tau_2 = \hat{Q}(r^*) = Q(1 - r^*) = Q(1 - (1 - r)) = Q(r) = \tau_1$$

Thus the above two statements have the same validity. Thus we see that the two statements

Most senior employees have ***high*** salaries

has the same validity as the statement

Few senior employees have ***not high*** salaries

3 Information Content of Linguistic Summaries

In the preceding we have introduced the idea of linguistic summaries and described a technique for determining their validity. One purpose in providing linguistic summaries is to provide useful global information about the database. In determining the usefulness of a summary a reasonable measure is some indication of the amount of information conveyed by the summary. At first impulse it is natural to think that only the degree of validity of a summary indicates the information about usefulness of a summary. As the following situation illustrates this is not the case. Assume we have a database of employees and consider the summary

Most employees are over 10 years old.

The fact that this is valid really doesn't convey much information. As a matter of fact this is manifestation of the following observation about the measure of validity. Consider two summaries

Q objects are S_1
Q objects are S_2

where Q is a monotonically increasing quantifier. It can be easily shown that if $S_1 \subseteq S_2$, $S_1(x) \leq S_2(x)$ for all x, then $\tau_2 \geq \tau_1$. Thus for monotonic increasing quantifiers we can always increase the validity of a summary by broadening the summarizer used in the summary. Similarly if Q_1 and Q_2 are two monotonically increasing quantifiers such that $Q_1 \subseteq Q_2$ then for a fixed S, $\tau(Q_1, S) \leq \tau(Q_2, S)$. For example the summary "at least 50% of the people are tall" will have a smaller degree of validity then the summary "at least 25% of the people are tall." We see that if we make the quantifier or summarizer to broad, while we can increase the validity. By broadening to much we can reach a point where the content of the summary is vacuous. Thus the transmission of useful information by linguistic summaries requires some trade-off between the size of the quantifiers and summarizer used and the resulting validity.

In order to more formally discuss the idea of informativeness associated with a linguistic summary we introduce some further concepts from fuzzy set theory. Assume V is some attribute with associated the domain X. Let $F_1, F_2, \ldots, F_q$ be a collection of fuzzy subsets corresponding to linguistic concepts associated with the attribute. In [35] Yager discusses the idea of the specificity of a fuzzy subset. Essentially the specificity measures the degree to which the fuzzy subsets points to one element as the manifestation of that fuzzy subset. For example the concept 30 years old is more specific then "about thirty" which in terms is more specific then "at least 20." In [36] measures of specificity associated with a normal fuzzy subset over the space X were introduced. Here we shall use one of these.

Assume the domain of V is the interval $X = [a, b]$ and let F be a normal fuzzy subset defined over X. Then the specificity is $Sp(F) = 1 - \frac{1}{b-a}\int_a^b F(x)dx$. Thus in this case the it is the negation of the average membership grade.

In the case when X is a finite set $X = \{x_1, x_2, \ldots, x_n\}$ then $Sp(F) = 1 + \frac{1}{n} - \frac{1}{n-1}\sum_{i=1}^{n} F(x_i)$ this is essentially the negation of the average of the membership grades.

It can be easily shown that the measure of specificity has the following properties

1. $Sp(F) = 1$ if $F(x) = 1$ for exactly one element in X and $F(x) = 0$ for all other elements
2. $Sp(F) = 0$ if $F = X$
3. $Sp(F_1) \geq Sp(F_2)$ if $F_1 \subset F_2$.

In the case of linguistic quantifiers since $X = [0, 1]$ we get $Sp(Q) = 1 - \int_0^1 Q(x)dx$, the negation of the area of the quantifier. We further note that if $\bar{Q}$ is the negation of Q then $Sp(\bar{Q}) = 1 - \int_0^1 1 - Q(x)dx = \int_0^1 Q(x)dx$ hence $Sp(\bar{Q}) = 1 - Sp(Q)$. On the other hand if $\hat{Q}$ is the antonym of Q, $\hat{Q}(r) = Q(1-r)$, we get that $Sp(\hat{Q}) = 1 - \int_0^1 \hat{Q}(x)dx = 1 - \int_0^1 Q(1-x)dx$. Replacing $1-x$ by y we get $Sp(\hat{Q}) = 1 - \int_0^1 \hat{Q}(y)(-dy) = 1 - \int_0^1 Q(y)dy$ hence

$Sp(Q) = Sp(\hat{Q})$. Thus the specificity of the antonym of a quantifier is the same as the specificity of the quantifier while the negation has a specificity equal to the complement.

We look at the specificity of a number of prototypical quantifiers. We start with non-decreasing quantifiers. Consider the quantifier "at least α", $Q(x) = 0$ if $x < \alpha$ and $Q(x) = 1$ if $x > \alpha$. In this case $Sp(Q) = 1 - \int_0^1 Q(y)dy = 1 - \int_\alpha^1 dx = 1 - (1 - \alpha) = \alpha$. Consider the quantifier $Q(r) = r^\beta$, where $\beta > 0$, here $Sp(Q) = 1 - \int_0^1 r^\beta dx = 1 - \frac{1}{\beta+1} = \frac{\beta}{1+\beta}$. The specificity increases as β increases.

Previously we defined "most" as $Q(r) = 0$ if $r < 0.5$ and $Q(r) = (2r-1)^{1/2}$ if $r \geq 0.5$. For this quantifier we get $Sp(Q) = 1 - \int_{0.5}^1 (2r-1)^{1/2} dr = 2/3$. More generally we can consider the class $Q(r) = 0$ if $r < \alpha$ and $Q(r) = (\frac{r-\alpha}{1-\alpha})^\beta$ if $r \geq \alpha$, we assume $\alpha \leq 1$ and $\beta \geq 0$. In this case

$$Sp(Q) = \frac{\beta + \alpha}{\beta + 1}.$$

We note if $\beta = 0$ we get the first case and if $\alpha = 0$ we get the second case. To some extend $Sp(Q)$ provides a crisp approximation of the quantity in agreement. We can refer to this as the focus of the quantifier.

Let us now consider the decreasing quantifiers, those for which $Q(r_1) \geq Q(r_2)$ if $r_1 < r_2$. We recall that if Q is a decreasing quantifier then its antonym $\hat{Q}$, where $\hat{Q}(r) = Q(1 - r)$ is a nondecreasing quantifier. Thus we see that the non-increasing and nondecreasing always correspond to antonym pairs. Furthermore since $Sp(Q) = Sp(\hat{Q})$ these pairs attain the same specificity. In particular we note that "at most α" has as its antonym "at least $(1 - \alpha)$" which has specificity $1 - \alpha$.

Finally we consider unimodal quantifiers. One important class of unimodal quantifiers are those centered about some value a and having spread b, $Q(r) = 0$ if $0 \leq r \leq a-b$, $Q(r) = 1$ if $a-b \leq r \leq a+b$ and $Q(r) = 0$ if $r \geq a+b$, where we assume $a-b \geq 0$ and $a+b \leq 1$. In this case $Sp(Q) = 1 - \int_{a-b}^{a+b} dx = 1 - 2b$. We note specificity is indifferent to value a the focus of the quantifier. Here we introduce the general idea of the focus of a unimodal quantifier. If Q is a unimodal we can define the focus as $FOC(Q) = \frac{\int_0^1 Q(x)x dx}{\int_0^1 Q(x)dx}$. We can suggest an alternative definition. Let Q be a unimodal quantifier with $Q(r) = 1$ for $r \in [a, b]$. Q is monotonically non- decreasing for $r < a$ and Q is monotonically non-increasing for $r > b$. We define Q^* as $Q^*(r) = Q(r)$ for $r \leq a$ and $Q^*(r) = 1$ for $r > a$. Q^* is monotonically non-decreasing. We define Q_* as $Q_*(r) = Q(r)$ for $r \leq b$ and $Q_*(r) = 1$ for $r < b$. Q_* is monotonically non-increasing. We then define $Foc(Q) = \frac{1}{2}(Sp(Q^*) + (1 - Sp(Q_*))) = \frac{1}{2} + \frac{1}{2}(Sp(Q^*) - Sp(Q_*))$.

Having introduced the idea of specificity we are now in a position to discuss the measure of information associated with a linguistic summary.

Consider a typical simple linguistic summary involving a nondecreasing quantifier,

$$Q_1 \text{ objects have } V \text{ is } S_1$$

Assume that with respect to our database we can establish this summary has validity equal to τ_1. This type of statement provides more useful information if the validity is large. In addition informativeness is increased if S_1 is a narrow fuzzy subset, the specificity of S_1 is large. With respect to Q_1 we prefer a higher the focus, a large specificity which corresponds to a narrow fuzzy subset. Using this observation we can provide as a measure of useful information of this type of linguistic summary

$$I(Q_1, S_1) = \tau_1 Sp(Q_1) Sp(S_1)$$

Let us now turn to the case in which we have a non-increasing quantifier,

$$Q_2 \text{ objects have } V \text{ is } S_2$$

which has validity τ_2. We have previously shown that this is equivalent to the following summary

$$\hat{Q}_2 \text{ objects have } V \text{ is } \bar{S}_2$$

where $\hat{Q}_2$ is the antonym of Q_2. Since with Q_2 a non-increasing quantifier we have $\hat{Q}_2$ is non-decreasing and hence we can express our measure of usefulness of summary using the preceding form

$$I(Q_2, S_2) = \tau_2 Sp(\hat{Q}_2) Sp(\bar{S}_2).$$

Furthermore we have shown that for antonym

$$Sp(\hat{Q}_2) = Sp(Q_2)$$

thus we can express the information as

$$I(Q_2, S_2) = \tau_2 Sp(Q_2) Sp(\bar{S}_2).$$

We now turn to unimodal statements

$$Q_3 \text{ objects have } V \text{ is } S_3$$

with validity τ_3. The situation here is somewhat more complex then in the preceding cases. Again we note that informative is increased by making Q_3 and S_3 specific and τ_3 large. Thus one form is

$$I(Q_2, S_2) = \tau_3 Sp(Q_3) Sp(S_2)$$

However there is one other consideration, location. Consider the two summaries

About 75% of the objects are $\underline{t}$all
About 25% of the objects are $\underline{t}$all.

We can see that the first statement has provided more useful information. In order to introduce this aspect we must use the idea of the focus of the quantifier. Using this we suggest as a measure of usefulness of a unimodal quantifier

$$I(Q_3, S_3) = Foc(Q_3)Sp(Q_3)\tau Sp(s)]$$

We shall not pursue the situation of unimodal quantifiers further.

Let us now turn to the informativeness of summaries involving association rules, those of the type Q *tall people are young.* Formally we can represent this as **QR are S**

First we shall consider the case in which Q is a non-decreasing quantifier. Again informativeness is improved if τ is large and Q and S are specific. We must consider the effect of R on informativeness. Consider the two propositions

All twenty year olds are tall
All people are tall

We easily see that the second summary has provided more information. More generally we can see that the broader the context of the antecedent, everything else being equal, the more informative the summary. Thus we suggest for the measure of informativeness of

$$Q_1 R_1 \text{ are } S_1$$

where Q_1 is non-decreasing.

$$I(Q_1, R_1, S_1) = \tau_1 Sp(Q_1)Sp(S_1)(1 - Sp(R_1))$$

Thus here we want wide antecedents, narrow consequents and large quantities in agreement as well as large validity

It is interesting to note that if R_1 is the whole space then the above statement becomes QX are S_1 an unqualified summary. Here $R = -X$, hence the specificity of R is $Sp(X) = 0$ thus we get $(1 - Sp(R)) = 1$ and hence our suggested measure reduces to our unqualified measure.

For qualified summaries involving non-increasing or unimodal quantifiers we extend the measure of informativeness by adding the term $(1 - Sp(R))$ as in the case of qualified summaries involving non-decreasing quantifiers.

4 Data Mining Using Summaries

We now turn to the use of these linguistic summaries in data mining. The structure of a linguistic summary provides a very natural framework for making statements about the global knowledge in a database. In the preceding we have described a mechanism for determining the validity of linguistic summaries based upon their consistency with the data in the database. We have also suggested a mechanism for measuring the usefulness of the information contained in a linguistic summary. In order to be able to use these linguistic summaries for database mining one piece is still missing. Since there are many possible summaries available we need some indication of the linguistic summaries we should try to put to our test of validity. This problem can be seen as part of the more general issue of **conjecturing**. The problem of conjecturing can be seen as very much at the heart of the problem of data mining. In order to make some conjectures about the global knowledge content in the database we need some direction.

One place to look for direction is to the language people use to discuss an attribute that appear in the database. Consider the attribute age. There are special concepts and words that people use in making statements about age. Terms like *young*, *old*, *teen* play an important role in discussing age. In one approach to data mining we can make use of these concepts to help provide direction for our conjecturing problem. Essentially we are saying that there are certain concepts that human beings have found in the past to be helpful in expressing knowledge about this attribute and let us initially look to these concepts to help formulate our linguistic summaries. We call techniques based upon using user supplied language **template** based discovery methods.

In the template based method we are initially looking outside the database for some direction to help make conjectures. Another place to look for direction is within the database itself. Using this technique we can let the data provide some direction. We shall call techniques based upon this approach **clustering** based discovery methods.

In the following we shall discuss in turn these two approaches to data discovery.

5 Template Based Data Mining

In this section we describe the basic ideas of the template approach to data mining. Assume V is some attribute of a database with domain X. A template for this attribute is a set of linguistic concepts associated with this attribute. Typically elements in this template set are linguistic terms useful for describing in a granular way this attribute. Each object in the set is called a template term and is represented by a fuzzy subset of X. We shall denote this template set as

$$Temp_V(X) = \{S_1, S_2, \ldots, S_{n_v}\}$$

Generally the template set should at the very least provide a fuzzy partitioning of the domain of the attribute variable. For example if V is the attribute age than a the template set could be {young, middle age, old}. Using this template set as our primary vocabulary we can, if needed, construct other linguistic terms by forming terms using fuzzy set operations such union, intersection, negation and intensification. [37]

The actual choice and semantics of the template set must be determined within the context of database we are investigating and the goals of the user. We note that for each attribute in the database we can provide such a template.

In addition we must have a template set of linguistic quantifiers

$$Temp_Q = \{Q_1, Q_2, \dots, Q_p\}.$$

Each element in $Temp_Q$ is a pair consisting of a linguistic term, quantifier, along with a fuzzy subset defining that quantifier. These objects will provide the terms used to express the quantities in agreement in our linguistic summaries. Generally we shall choose the elements in $Temp_Q$ to be a fuzzy partitioning of the unit interval. Implicit in this partitioning is an ordering of the quantified values, in terms of increasing value. We shall assume under this ordering $Q_i > Q_j$ if $i > j$. It will also be useful if each of the terms in $Temp_Q$ has about the same width, specificity. Another useful tool will be a mapping M which associates with each $r \in I$ a quantifier Q^* which has the largest membership grade for that value, $M(r) = Q^*$ where $Q^*(r) = \text{Max}_i Q_i(r)$. In addition for each quantifier Q_i we can calculate its centroid value,

$$Cent(Q_i) = \frac{\int_0^1 Q_i(x)\ x\ dx}{\int_0^1 Q_i(x)\ dx}$$

We are now in the position to begin the process of data mining. Let $D = \{x_1, x_2, \dots, x_m\}$ indicate the set of objects that exists in our database. Assume V is our attribute of interest. For each $S_i \in Temp_V$ we calculate $v_i = \sum_{k=1}^{m} S_i(x_k)$ and $r_i = \frac{v_i}{m}$.

We then calculate for each S_i the value of the quantifier Q_i^* best associated with r_i, that is $Q_i^* = M(r_i)$. At this stage we have a collection of pairs, one for each S_i in $Temp_V$. Each pair (S_i, Q_i^*) consists of a template value S_i and its corresponding quantifier. Each of these pairs corresponds to a valid summary of the form

$$Q_i^* \text{ elements have } V \text{ as } S_i$$

Using these pairs we are in the position to begin providing global knowledge about this attribute. Let us look at the different types of knowledge we can provide. First we can calculate for each pair its useful information content $I(S_i, Q_i^*)$. We can then provide the user with the summaries having the largest information content, the most informative. In addition we can select the linguistic summary with the largest Q^* value and try to improve the information associated with it. For example assume S_4 has the largest Q^* value, Q_4. We can now introduce a modified template, $S_a = (S_4)^2, S_a(y) = (S_4(y))^2$ which corresponds to very S_4. We can calculate $r_a = \frac{\sum_{j=1}^{m} S_a(x_j)}{m}$ and $Q_a^* = M(r_a)$. This gives us a new linguistic summary (S_a, Q_a^*). If the information content of this is high compared with the original summaries this could be useful. We could also modify S_4 by softening it, formulating $S_b = (S_4)^{1/2}$ which corresponds to sort of S_4. Again we can calculate the Q value associated with this and see if it increases informativeness.

One can obtain new meta-knowledge by combining neighboring summarizers by taking their union. Another direction one can look to for providing meta-knowledge about this attribute involves the use of the centroids of the Q_i^*. For the quantifier associated with a S_i calculate its centroid $Cent(Q_i^*) = C_i$. Let $u_i = \frac{C_i}{\sum_{j=1}^n C_j}$. Let $H_v = -\sum_{i=1}^n u_i \ln(u_i)$, which is an entropy like measure associated with the template set. If H_v is close to $\ln n$ then we have that the values for V are uniformally distributed with respect to the template set. If H_v is close to zero than we can ascertain that the distribution of values of V has a very strong peak at the template having the largest quantifier.

We now turn to the generation of linguistic summaries involving multiple attributes. Assume V and U are two attributes and let $\{S_1, S_2, \ldots, S_{n_v}\}$ and $\{R_1, R_2, \ldots, R_{n_u}\}$ be their respective template sets. For each pair S_i and R_j we calculate

$$v_{ij} = \sum_{k=1}^{m} S_i(x_k) R_j(x_k)$$

and

$$r_{ij} = \frac{v_{ij}}{m}.$$

We then calculate for r_{ij} the quantifier Q_{ij}^* that is best associated with r_{ij}. This gives us a collection of triples (Q_{ij}^*, S_i, R_j). Each of these corresponds to a valid linguistic summary of the form

Q_{ij}^* elements have V is S_i and U is R_j.

Again for each of these we can calculate the degree of informative and the report to a user the summary with the best information context.

We can of course use the above method to combine multiple number of attributes. In this case to avoid computational explosion we may just desire to combine template values that have high quantities of elements.

Another type of proposition involving more than one attribute are those of the form

if V is S_i then Q U are R_j.

To find these kinds of summaries we proceed as follows. For each S_i we calculate

$$\hat{v}_{ij} = \sum_{k=1}^{m} S_i(x_k) R_j(x_k)$$

and

$$\hat{r}_{ij} = \frac{v_{ij}}{m} \qquad \text{for } j = 1, \ldots n_u$$

We then calculate for each $\hat{r}_{ij}$ the quantifier $\hat{Q}_{ij}$ that it is best associated with. This gives us a collection of tuples $(\hat{Q}_{ij}, S_i, R_j)$. Each of these corresponds to a valid linguistic summary of the form

If V is S_j then $\hat{Q}_{ij}$ U are R_j.

6 Data Mining with Mountain Clustering

We now turn to data driven clustering methods. In [27–29] Yager and Filev introduced a clustering method which they called the mountain method. We first describe this method and then indicate its role in the problem of database mining.

Let $A_1, A_2, \ldots, A_m$ be a set of attributes which take their values in the spaces X_i respectively. We shall let $X = X_1 \times X_2 \cdots \times X_m$ be our feature space. Furthermore we shall assume that there exists on this space some metric which allows us to measure the distance between objects in this space. In addition we assume that we have a collection $D =< d_1, d_2, \ldots d_n >$ of data from this space, in particular $d_i \in X$.

The mountain method is an algorithm used for the estimation of cluster centers associated with this data. As we shall see this method requires no predetermination of the number of cluster centers. The mountain method consists of three basic steps.

1. Selection of a finite subset of elements E from the feature space X to be considered as potential cluster centers. We call this the granularization step.
2. Introduction of the data and <u>construction</u> of the mountain function M defined on E.
3. Selection of cluster centers by <u>destruction</u> of the mountain function.

We shall now look in more detail at each of these steps. We first start with the granularization step. Assume X is the feature space. In this step we select a finite subset E of X. This subset, whose elements are called the nodes, will form the set of <u>potential</u> cluster centers. Thus the discovered cluster centers will be in E. In selecting E we are faced with a basic trade-off which we must keep in mind. The larger the set E the more computationally intense the remaining steps. The smaller the set E the more approximate the discovered cluster centers. A number of techniques can be suggested for the selection of the subset E. A first technique is to randomly select from X some fixed number of points. A second approach, the one originally suggested in [28], is to provide a griding of the space X and select as our set E the intersection of these grids. This approach is illustrated in Figure 2 for the two dimension case.

In figure 2 the heavy dots, which are at the intersection of our grid lines, provide the set of elements making up the space E. It should be noted that the griding need not be uniform and could be more or less dense in different parts of the space.

Another approach, suggested by Chui [38], is to use the data set D to provide the set E. In using this method the points in E are the distinct points in the bag D. Another approach which we shall just mention is a knowledge based approach. Here we used any available additional information we may have to select the points in E.

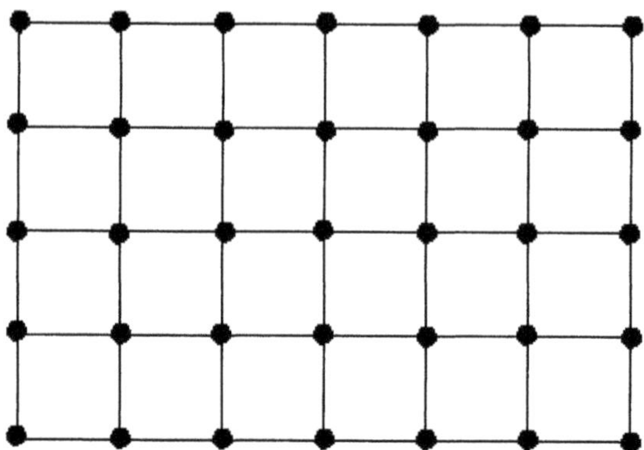

Fig. 2. Grid for Mountain Method

Whatever method we use, as a result of the granularization step, we have a finite set E of potential cluster centers. The next step is the construction of the mountain function. As we noted the mountain function M is a function defined over the space E of potential cluster centers. We construct the mountain function from the data set $D =< d_1, d_2, \ldots, d_n >$. In particular for each point $z_i \in E$ we form

$$M(z_i) = \sum_{k=1}^{n} e^{-\rho(z_i, d_k)}$$

where $\rho(z_i, d_k)$ is some metric related to the distance between z_i and d_k. Thus at each point z_i, the mountain function is a reflection of the density of the data points in its neighborhood.

In the final step of the mountain method the cluster centers are obtained by a destruction of the mountain function. Assume that the maximal value of the mountain function M is M_1^* and occurs at the point p_1, that is

$$M_1^* = M(p_1) = \text{Max}_i[M(z_i)].$$

In this case p_1 becomes our first cluster center. We next from a revised mountain function $\hat{M}_2$ by eliminating the effects of the selected cluster center,

$$\hat{M}_2(z) = M(z) - M_1^* e^{-\rho(z, p_1)}$$

This operation corresponds to a destruction of the mountain function. This new revised mountain function, $\hat{M}_2$, is now used to find the second cluster center. Assume the maximal value, M_2^*, of the revised mountain function occurs at the node p_2. This becomes our second cluster center. We now remove the effect of this second cluster center to find the remaining centers. More generally, if p_k is the location of the kth cluster center, the point of occurrence of the maximal value of $\hat{M}_k$, and if $\hat{M}_K(p_k) = M_K^*$ then we form the revised mountain function

$$\hat{M}_{k+1}(z) = \hat{M}_k(z) - M_K^* e^{-\rho(z,p_k)}$$

We continue in this manner until we virtually destroy the mountain function. As a result of the application of the mountain method we obtain a collection, $p_1, p_2, \ldots, p_q$, of points which will serve as the basis of the formulation of linguistic summaries about our database. At this point we can construct a collection of fuzzy subsets $\{\tilde{P}_1, \ldots, \tilde{P}_q\}$ each $\tilde{P}_j$ corresponds to the concept "about p_j". We then treat this collection of fuzzy subsets in the same manner as we treated the collection of template values in the previously section. Here then we generate linguistic summaries such as Q objects in the database are "about p_j". We can also generate other summarizes using other hedges associated with the p_j nuclei, for example moderately close to p_j. very close to p_j etc.

7 Conclusion

In the preceding we have described an approach to discovery in databases using fuzzy set techniques. Central to this approach is the idea of a linguistic summary. We suggested two methods for addressing the conjecturing problem, the template method and the mountain clustering method.

References

1. Piatetsky-Shapiro, G. and Frawley, B., Knowledge Discovery in Databases, MIT Press: Cambridge, MA, 1991.
2. Piatetsky-Shapiro, G., Knowledge Discovery in Databases: Papers from the 1993 AAAI Workshop, AAAI Press: Menlo Park, CA, 1993.
3. Fayyad, U. M. and Uthurusamy, R., Knowledge Discovery in Databases, AAAI Press: Menlo Park, Ca, 1994.
4. Fayyad, U. M. and Uthurusamy, R., "Proceedings of the First International Conference on Knowledge Discovery and Data Mining," AAAI Press, Menlo Park, CA, 1995.
5. Matheus, C. J., Chan, P. K. and Piatetsky-Shapiro, G., "Systems for knowledge discovery in databases," IEEE Transactions on Knowledge and Data Engineering 5, 903-913, 1993.
6. Adriaans, P. and Zantinge, D., Data Mining, Addison-Wesley: Reading, MA, 1996.
7. Fayyad, U., Piatetsky-Shapiro, G., Smyth, P. and Uthurusamy, R., Advances in Knowledge Discovery and Data Mining, AAAI Press: Menlo Park, Ca, 1996.

8. Berry, M. J. A. and Linoff, G., Data Mining Techniques, John Wiley & Sons: New York, 1997.
9. Han, J. and Kamber, M., Data Mining: Concepts and Techniques, Morgan Kaufmann: San Francisco, 2001.
10. Lin, T. S., Yao, Y. Y. and Zadeh, L. A., Data Mining, Rough Sets and Granular Computing, Physica-Verlag: Heidelberg, 2002.
11. Borgelt, C. and Kruse, R., Graphical Models: Methods for Data Analysis and Mining, John Wiley & Sons: New York, 2002.
12. Mena, J., Data Mining Your Website, Digital Press: 1999.
13. Linoff, G. S. and Berry, M. J. A., Mining the Web, John Wiley & Sons: New York, 2001.
14. Yager, R. R., "On linguistic summaries of data," in Knowledge Discovery in Databases,Piatetsky-Shapiro, G. & Frawley, B. (eds.), Cambridge, MA: MIT Press, 347-363, 1991.
15. Kacprzyk, J. and Strykowski, P., "Linguistic data summaries for intelligent decision support," in Fuzzy Decision Analysis and Recognition Technology, edited by Felix, R., EDFAN: 3-12, 1999.
16. Kacprzyk, J. and Zadrozny, S., "Data mining via fuzzy querying over the Internet," in Knowledge Management in Fuzzy Databases, edited by Pons, O., Vila, M. A. and J., K., Physica-Verlag: Heidelberg, 211-233, 2000.
17. Kacprzyk, J., Yager, R. R. and Zadrozny, S., "A fuzzy logic based approach to linguistic summaries in databases," International Journal of Applied Mathematical Computer Science 10, 813-834, 2000.
18. Kacprzyk, J., Yager, R. R. and Zadrozny, S., "Fuzzy linguistic summaries of databases for efficient business data analysis and decision support," in Knowledge Discovery for Business Information Systems, edited by Abramowicz, W. and Zaruda, J., Kluwer Academic Publishers: Hingham, MA, 129-152, 2001.
19. Kacprzyk, J. and Yager, R. R., "Linguistic summaries of data using fuzzy logic," International Journal of General Systems 30, 133-154, 2001.
20. Rasmussen, D. and Yager, R. R., "Using summarySQL as a tool for finding fuzzy and gradual functional dependencies," Proceedings of the Sixth International Conference on Management of Uncertainty in Knowledge-Based Systems, Granada, 275-280, 1996.
21. Rasmussen, D. and Yager, R. R., "A fuzzy SQL summary language for data discovery," in Fuzzy Information Engineering: A Guided Tour of Applications, edited by Dubois, D., Prade, H. and Yager, R. R., John Wiley & Sons: New York, 253-264, 1997.
22. Rasmussen, D. and Yager, R. R., "SummarySQL-A fuzzy tool for data mining," Intelligent Data Analysis-An International Journal 1 (Electronic Publication), URL-http:www- east.elsevier.com/ida/browse/96-6/ida96-6.htm, 1997.
23. Rasmussen, D. and Yager, R. R., "Finding fuzzy and gradual functional dependencies with summarySQL," Fuzzy Sets and Systems 106, 131-142, 1999.
24. Yager, R. R. and Kacprzyk, J., "Linguistic data summaries: a perspective," Proceedings of the Eight International Fuzzy Systems Association World Congress, Taiwan Vol 1, 44-48, 1999.
25. Zadeh, L. A., "Fuzzy logic = computing with words," IEEE Transactions on Fuzzy Systems 4, 103-111, 1996.
26. Zadeh, L. A., "Outline of a computational theory of perceptions based on computing with words," in Soft Computing and Intelligent Systems, edited by Sinha, N. K. and Gupta, M. M., Academic Press: Boston, 3-22, 1999.

27. Yager, R. R. and Filev, D. P., "Learning of fuzzy rules by mountain clustering," SPIE Conference on Applications of Fuzzy Logic Technology, Boston, 246-254, 1993.
28. Yager, R. R. and Filev, D. P., "Approximate clustering via the mountain method," IEEE Transactions on Systems, Man and Cybernetics 24, 1279-1284, 1994.
29. Yager, R. R. and Filev, D. P., "Generation of fuzzy rules by mountain clustering," Journal of Intelligent and Fuzzy Systems 2, 209-219, 1994.
30. Zadeh, L. A., "A computational approach to fuzzy quantifiers in natural languages," Computing and Mathematics with Applications 9, 149-184, 1983.
31. Yager, R. R., "Quantifiers in the formulation of multiple objective decision functions," Information Sciences 31, 107-139, 1983.
32. Yager, R. R., "Reasoning with fuzzy quantified statements: part I," Kybernetes 14, 233-240, 1985.
33. Yager, R. R., "Reasoning with fuzzy quantified statements: part II," Kybernetes 15, 111- 120, 1986.
34. Yager, R. R., "Database discovery using fuzzy sets," International Journal of Intelligent Systems 11, 691-712, 1996.
35. Yager, R. R., "On the specificity of a possibility distribution," Fuzzy Sets and Systems 50, 279-292, 1992.
36. Yager, R. R., "Default knowledge and measures of specificity," Information Sciences 61, 1- 44, 1992.
37. Zadeh, L. A., "A theory of approximate reasoning," in Machine Intelligence, Vol. 9, edited by Hayes, J., Michie, D. and Mikulich, L. I., Halstead Press: New York, 149-194, 1979.
38. Chui, S. L., "Fuzzy model identification based on cluster identification," Journal of Intelligent and Fuzzy Systems 2, 267-278, 1994.

Index

GPSR Compliance
The European Union's (EU) General Product Safety Regulation (GPSR) is a set of rules that requires consumer products to be safe and our obligations to ensure this.

If you have any concerns about our products, you can contact us on

ProductSafety@springernature.com

In case Publisher is established outside the EU, the EU authorized representative is:

Springer Nature Customer Service Center GmbH
Europaplatz 3
69115 Heidelberg, Germany

www.ingramcontent.com/pod-product-compliance
Ingram Content Group UK Ltd.
Pitfield, Milton Keynes, MK11 3LW, UK
UKHW021828190726
13853UKWH00003B/1254
* 9 7 8 3 6 4 2 5 3 5 2 8 4 *